Wastewater Biology: The Life Processes

A Special Publication

Prepared by **Task Force on Wastewater Biology: The Life Processes**
Michael H. Gerardi, *Chair*

Robert M. Arthur
Khalil Z. Atasi
Mark T. Baumgardner
Jerry T. Cheshuk
Steven Chiesa
Koby T. Crabtree*
C.R. Curds
Allan J. DeLorme
Bhupinder Dhaliwal
Martha Anne Dow
Ronald Elks
Jerzy J. Ganczarczyk
Anthony Gaudy, Jr.
Joseph J. Gauthier
James Grimm
Joanne L. Grimm

Roger R. Hlavek
Leeann Johnson
Donna Kaluzniak
Richard B. Kapuscinski
Kenneth E. Kaszubowski
P.S. Kodukula
Hubert A. Lechevalier
Audrey D. Levine
Dennis R. Lindeke
Charles L. Logue
Miryoussef Norouzian
Robert E. Pariseau
Malcolm D. Peacock
Raymond W. Regan
Frank Rogalla
Ronald G. Schuyler

Sophie G. Simon
Thomas J. Simpkin
Edward V. Skonieczny
Jeffrey C. Steven
Thomas L. Stokes, Jr.
Penny K. Stowe
R. Yucel Tokuz
Allan R. Townshend
Joseph C. Traurig
James M. Turek
David A. Vaccari
Scott E. Walters
Randel L. West
Robert C. Wichser
Melvin C. Zimmerman

* Deceased

T0155443

Under the Direction of the **Operations and Maintenance Subcommittee** of the **Technical Practice Committee**

Water Environment Federation
601 Wythe Street
Alexandria, VA 22314–1994 USA

Abstract

This publication is a continuation of the effort to develop *Wastewater Biology: The Microlife*, which was published by the Water Environment Federation in 1990. *Wastewater Biology: The Microlife* describes the ecology and beneficial and detrimental roles of the microscopic life forms—the microlife—found in wastewater treatment processes.

Wastewater Biology: The Life Processes provides an introduction and basic comprehensive review of the life processes in wastewater treatment systems that are responsible for the degradation of wastes and the production of acceptable effluents.

The book reviews organisms, their metabolic schemes used in the degradation of wastes, and environmental factors that positively and negatively affect the life processes. Written with a minimum of technical jargon, wastewater personnel do not need a formal background in biology or microbiology to use the book.

The publication should provide operators everyday guidance needed to monitor and regulate treatment processes in wastewater treatment plants. The book also provides wastewater personnel such as administrators, biologists, chemists, engineers, and technicians with a reference, an operational tool, and an educational resource about the life processes.

Chapter 1 describes an overview of the book and provides guidance to its use. Chapter 2 reviews how wastes are degradable by bacteria through the use of enzyme systems and the way enzyme systems are affected by the wastewater environment. Chapter 3 describes bacterial nutrient requirements necessary to maintain an active and healthy biomass capable of producing an acceptable effluent quality. Chapter 4 describes the role of heterotrophic and autotrophic bacteria in wastewater treatment processes. Chapter 5 discusses volatile acids and methanogenic bacteria, and Chapter 6 reviews metabolic processes of bacteria in wastewater. Chapter 7 describes the process of dealing with heavy metals in wastewater treatment, and Chapter 8 discusses sources and effects of organic compounds in wastewater. Chapter 9 covers the way bioaugmentation or biomass enhancement is used to supplement enzyme activity in a wastewater treatment process.

The Water Environment Federation is a nonprofit, educational organization composed of member and affiliated associations throughout the world. Since 1928, WEF has represented water quality specialists, including biologists, bacteriologists, local and national government officials, treatment plant operators, laboratory technicians, chemists, industrial technologists, students, academics, equipment manufacturers/distributors, and civil, design, and environmental engineers.

For information on membership, publications, and conferences, contact

Water Environment Federation
601 Wythe Street
Alexandria, VA 22314–1994 USA
(703) 684-2400

Library of Congress Cataloguing-in-Publication Data

Wastewater biology. The life processes: a special publication / prepared by
    Task Force on Wastewater Biology: the Life Processes under the direction
    of the Operations and Maintenance Subcommittee of the Technical Prac-
    tice Committee.
        p. cm.
    Includes bibliographical references and index.
    ISBN 1-881369-93-5
        1. Sewage—Purification—Biological treatment.   2. Sewage—Microbiol-
    ogy. I. Water Environment Federation. Task Force on Wastewater Biology:
    the Life Processes.   II. Title: Life processes.
    TD755.W29   94-37376
    628.3'5—dc20   CIP

Library of Congress Catalog No.
ISBN 1-881369-93-5
Printed in the USA

Special Publications of the Water Environment Federation

The WEF Technical Practice Committee (formerly the Committee on Sewage and Industrial Wastes Practice of the Federation of Sewage and Industrial Wastes Associations) was created by the Federation Board of Control on October 11, 1941. The primary function of the Committee is to originate and produce, through appropriate subcommittees, special publications dealing with technical aspects of the broad interests of the Federation. These manuals are intended to provide background information through a review of technical practices and detailed procedures that research and experience have shown to be functional and practical.

IMPORTANT NOTICE

The contents of this publication are for general information only and are not intended to be a standard of the Water Environment Federation (WEF).

No reference made in this publication to any specific method, product, process, or service constitutes or implies an endorsement, recommendation, or warranty thereof by the Federation.

The Federation makes no representation or warranty of any kind, whether expressed or implied, concerning the accuracy, product, or process discussed in this publication and assumes no liability.

Anyone using this information assumes all liability arising from such use, including but not limited to infringement of any patent or patents.

Water Environment Federation Technical Practice Committee Control Group

F.D. Munsey, *Chair*
L.J. Glueckstein, *Vice-Chair*

T.L. Krause
C.N. Lowery
T. Popowchak
J. Semon

Authorized for Publication by the Board of Control
Water Environment Federation

Quincalee Brown, *Executive Director*

Preface

Wastewater Biology: The Life Processes reviews

- Bacteria involved in wastewater and sludge stabilization,
- Metabolic or life processes involved in wastewater and sludge stabilization,
- Environmental conditions that promote and sustain favorable life processes or desired treatment efficiency,
- Conditions that adversely affect the life processes or needed treatment efficiency, and
- Operational measures needed to monitor, correct, or enhance the life processes (or treatment efficiency).

As its comparison volume on microlife, this special publication is intended for operators and technicians concerned with the daily control of their treatment plants. Its emphasis is on presenting an operator's guide to the biological aspects of wastewater treatment and it is written with a minimum of technical jargon.

This publication was produced under the direction of Michael H. Gerardi, *Chair.* The principal contributing authors were

Robert M. Arthur	Randolph Harrison	Mesut Sezgin
Koby T. Crabtree*	Richard B. Kapuscinski	Daniel P. Smith
Allan J. DeLorme	Prasad S. Kodukula	Joseph C. Traurig
Joseph J. Gauthier	Audrey D. Levine	Scott E. Walters
Michael H. Gerardi	Raymond W. Regan	

*Deceased

In addition to the Task Force Chair and Technical Practice Committee Control Group members, reviewers include

Tom Barron	Lewis E. Goyette	Brian F. McNamara
Arthur Carnrick	Randolph Harrison	Ana M. Palmer
Edwin E. Geldreich	Theresa Koschny	Daniel P. Smith

Authors' and reviewers' efforts were supported by the following organizations:

Arthur Technology, Fond du Lac, Wisconsin
Bio Systems Corporation, Roscoe, Illinois
Blue Heron Environmental Technologies, Athens, Ontario, Canada
British Museum of Natural History, London, England
CH2M Hill, Denver, Colorado
Cie Gale Des Eaux, Maisons Laffitte, France

City of Jacksonville Public Utilities Department, Jacksonville, Florida
City of Yorkton, Saskatchewan, Canada
Department of Environmental Resources, Harrisonburg, Pennsylvania
Greenville Utilities Commission, Greenville, North Carolina
Hampton Roads Sanitation District, Virginia Beach, Virginia
James Madison University, Harrisonburg, Virginia
Lycoming College, Williamsport, Pennsylvania
Madison Metro Sewerage District, Madison, Wisconsin
Marketing and Consulting Services, Inc., Roanoke, Virginia
Marshall Durbin Co., Jackson, Mississippi
McNamee Porter & Seeley, Inc., Detroit, Michigan
Metro Waste Control Commission, Hastings, Minnesota; St. Paul, Minnesota
Nu-Aqua Incorporated, Richlandtown, Pennsylvania
Parkhill Smith & Cooper, Inc., Lubbock, Texas
Rothberg Tamburini Winsor, Denver, Colorado
Rutgers State University, Morrisville, Vermont
Santa Clara University, Santa Clara, California
Sigma Environmental, Oak Creek, Wisconsin
Stevens Institute of Technology, Hoboken, New Jersey
The Upjohn Company, Kalamazoo, Michigan
University of Alabama, Birmingham, Alabama
University of Michigan, Ann Arbor, Michigan
University of Wisconsin, Wausau, Wisconsin
U.S. Environmental Protection Agency, Cincinnati, Ohio
Utah State University, Logan, Utah
Village of Minerva Park, Columbus, Ohio
Williamsport Sanitary Authority, Williamsport, Pennsylvania

Federation technical staff project management was provided by Berinda J. Ross; technical editorial assistance was provided by William N. Wolle.

Contents

List of Tables

List of Figures

Chapter 1

Introduction

OVERVIEW

Wastewater Biology: The Life Processes is a presentation of the biological activities—the life processes—performed by the microlife in the stabilization of wastewater and is intended to provide biological knowledge to those operators and technicians working to provide acceptable treatment conditions in the wastewater environment. The text also should be of value to numerous wastewater professionals, including sanitary engineers, chemists, microbiologists, and educators.

Wastewater Biology: The Life Processes is a follow-up to *Wastewater Biology: The Microlife* (WPCF, 1990), which presented the description, ecology, and beneficial and detrimental roles of the microscopic life forms—the microlife—found in wastewater treatment processes.

This publication contains a minimum of technical jargon so that it can be reviewed and absorbed by individuals with different professional backgrounds. Because the manual includes the *taxonomy* (the naming and classification) of the microlife commonly found in a wastewater treatment plant, a brief review of taxonomic principles as presented in *Wastewater Biology: The Microlife* is suggested.

Wastewater Microlife

Living organisms too small to be seen by the naked eye are called microorganisms or microscopic life forms—the microlife. Many different types of microorganisms inhabit wastewater, including bacteria, protozoa, and algae.

Microbial nutrition is the basis for much of our wastewater treatment technology. Heterotrophic organisms, such as most bacteria, use organic compounds as sources of carbon and energy. Such organisms help stabilize wastewater and degrade sludges. Autotrophic organisms, such as green algae, do not help stabilize organic wastes because they use carbon dioxide and other inorganic compounds as carbon sources.

The oxygen requirement for organisms also is important in the design of treatment processes. Aerobic organisms require dissolved oxygen to function actively, while anaerobic organisms function only in the absence of oxygen. Facultative anaerobes can grow in the presence or absence of oxygen.

Untreated Wastewater

Untreated wastewater is an ideal microbiological medium, as it is rich in the organic and inorganic nutrients needed for microbial growth (McKinney, 1962). Almost every kind of organism can be found in wastewater, whether introduced at its source or by infiltration from the soil. Because of the high biochemical oxygen demand (BOD) of untreated wastewater, it provides a strongly anaerobic environment. Anaerobic and facultatively anaerobic bacteria thrive under these conditions. Aerobic organisms, such as protozoa, fungi, and aerobic bacteria, survive by entering a dormant form, such as a spore or cyst. Some of the lower forms of protozoa and fungi are able to survive at the wastewater-air interface, where oxygen can be present.

An example of a microbial-induced wastewater problem occurs in the conveyance system. Crown corrosion can occur on the top section, or crown, of a concrete sewer (Gaudy and Gaudy, 1980, and McKinney, 1962) and happens when sulfate-reducing bacteria, such as *Desulfovibrio*, convert sulfates to hydrogen sulfide under anaerobic conditions as in the following equation:

$$SO_4^{2-} + fats \rightarrow H_2S \tag{1.1}$$

Under the same conditions, anaerobic fermentation (see Chapter 4) can produce acids, thereby lowering the wastewater's pH. Under acidic conditions, hydrogen sulfide is volatile and escapes to the air, from where it may be absorbed in condensation on the wall of the sewer pipe. Under the aerobic conditions present, hydrogen sulfide can be oxidized by *Thiobacillus* to sulfuric acid, according to the following equation:

$$H_2S + 2O_2 \rightarrow H_2SO_4 \qquad\qquad (1.2)$$

Sulfuric acid chemically attacks the concrete. The relationship among the chemical species involved in crown corrosion is shown in Figure 1.1.

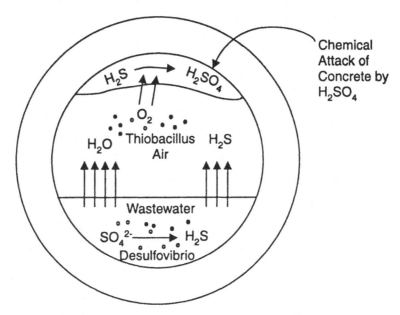

Figure 1.1   Chemical species involved in crown corrosion.

SECONDARY TREATMENT

TRICKLING FILTRATION. In trickling filters and other fixed-film systems, organisms grow in a film on a support (typically stone or plastic) and feed on organic matter found in wastewater (Cooke, 1967; Hammer, 1975; and McKinney, 1962). Although classified as aerobic treatment, both aerobic and anaerobic layers form in the biological film, as represented in Figure 1.2. Oxygen absorbed from the wastewater to the biological film is used by the aerobic organisms in the outer layer of the film. When the film is sufficiently thick, oxygen cannot reach the inner part near the support, so anaerobic and facultatively anaerobic bacteria predominate the inner layers.

While strictly aerobic bacteria such as *Bacillus* and *Zoogloea* are found in the outer aerobic layer and strictly anaerobic bacteria such as *Desulfovibrio* are found in the inner anaerobic layer, the majority of bacteria are facultative anaerobes. These bacteria, which include *Micrococcus, Pseudomonas, Alcaligenes, Enterobacteriaceae*, and *Flavobacterium*, live aerobically until the biological film becomes thick enough, after which they live anaerobically in the inner layer.

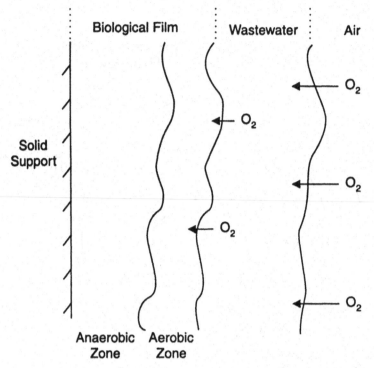

Figure 1.2   Aerobic and anaerobic regions in biological film of a trickling filter support.

Protozoa are also found on trickling filters, where they help remove the bacteria that feed on the organic wastes. The types of protozoa present will vary with changes in the food supply and from filter to filter.

Virtually all types of protozoa are found in filters. More *motile* members (organisms that have the ability to move spontaneously) predominate at the upper surface, where they can compete with bacteria for food, while nonmotile members are more prevalent in the lower regions. Motile protozoa include the flagellated types, such as *Euglena* and Sarcodina, pseudopod-propelled types, such as *Amoeba*, and the free-swimming ciliates, such as *Paramecium*. Nonmotile types include stalked ciliates, such as *Vorticella* and Suctoria, which attach themselves to the *substratum* (the solid of the filter support).

Many species of fungi have been identified on trickling filters, including species of *Aspergillus* and *Penicillium*. Fungi are strictly aerobic organisms and their growth usually is limited by their inability to compete with bacteria on a filter. However, under certain conditions, such as low pH, they can predominate over bacteria.

Other organisms, such as free-living nematodes and rotifers, live off the bacteria, protozoa, and fungi, contributing indirectly to the stabilization of organic wastes.

Algae need sunlight to survive and are found only at the upper level of the trickling filter. Algae resistant to organic pollution predominate, such as *Anacystis*, *Oscillatoria*, and *Phormidium*. Algal growth can clog the surface of the filter, especially in warm weather.

Filter flow rate determines the types of algae that predominate. On high-rate filters, diatoms and green algae are the main forms, while on low-rate filters, blue-green algae are plentiful.

ACTIVATED SLUDGE. Microscopic identification of the organisms found in activated sludge often is used in treatment plant operation because the effluent quality often can be correlated to the relative abundance, activity, and types of organisms present in the sludge (Dhaliwal, 1979; McKinney, 1962; Pike and Curds, 1971; Pipes, 1966 and 1967; Stowe, 1984; and U.S. EPA, 1977).

Bacteria are responsible for most organic waste stabilization in activated sludge. While many genera of bacteria have been identified in sludge, many believe that some, such as the intestinal genera *Escherichia* and *Klebsiella*, have little or no significance in the activated sludge process. Typically, *Achromobacter*, *Flavobacterium*, and *Pseudomonas* appear to be the dominant bacteria in floc formation. The presence of *Nitrosomonas* and *Nitrobacter* in activated sludge is responsible for nitrification, the conversion of ammonium (NH_4^+) to nitrite (NO_2^-) and nitrite to nitrate (NO_3^-).

Protozoa are the most useful type of organism for microscopic examination by wastewater treatment plant personnel in an activated sludge plant because they are easily observed and the quality of effluent being produced often can be indicated by the types of protozoa present.

The dominant type will depend on environmental and operational conditions. Sarcodina and flagellates predominate in activated sludge during periods of start-up or recovery from shock loads, as they are able to compete with bacteria for food at those times. Their dominance indicates a relatively low treatment efficiency.

However, as the number of bacteria increases, Sarcodina and flagellates can no longer effectively compete for food and the free-swimming ciliates become the dominant form of protozoa. As the free-swimming ciliates feed on bacteria, they act to reduce turbidity and further lower BOD in the effluent. As the wastes become stabilized, not enough food will remain to supply the high energy requirements of the free-swimming ciliates, and the crawling, stalked ciliates will become dominant because of their more efficient feeding mechanisms. A relative dominance of stalked ciliates indicates an efficient process.

Fungi also are present but not dominant in activated sludge, except in rare instances such as low pH or in specific industrial wastes. The genera *Geotrichum*, *Trichosporon*, *Penicillium*, *Cephalosporium*, and *Cladogorum* are common.

Rotifers and free-living nematodes are indicative of a well-stabilized system and a high-quality effluent. These more complex organisms are able to

---

digest larger pieces of organic materials than are bacteria and protozoa. Rotifer populations can become relatively large in extended aeration systems.

While most organisms efficiently remove organic waste in the activated sludge process, some can cause problems. For example, filamentous bacteria such as *Sphaerotilus natans* and fungi such as *Geotrichum* can cause sludge bulking. The filamentous actinomycete *Nocardia* has been shown to cause foaming problems.

OXIDATION PONDS. Many of the organisms found in oxidation ponds also are present in the other anaerobic treatment processes (Hawkes, 1983, and McKinney, 1962). A significant difference from the other processes is the existence of large algae populations in oxidation ponds. The types of organisms that dominate the oxidation pond depend on the pond's physical design and its loading.

As in trickling filters and activated sludge systems, aerobic and facultatively anaerobic bacteria are responsible for most of the organic waste stabilization. The dominant bacteria in activated sludge (*Pseudomonas*, *Flavobacterium*, and *Achromobacter*) also appear to be the dominant forms in an oxidation pond. A high mortality rate of coliform bacteria has been observed in oxidation ponds and has been attributed to the high pH produced by algal activity.

The other dominant organisms in oxidation ponds are algae. While the main function of algae in this waste stabilization process is to supply oxygen for other bacteria, there is evidence that some algae are capable of aiding in the stabilization process through chemosynthesis. They may be able to use simple organic molecules, such as the waste products of bacteria, to supplement carbon dioxide as a source of carbon.

Plant nutrients, especially nitrogen and phosphorus, are used by algae. The removal of these nutrients makes oxidation ponds more effective in controlling *eutrophication* (nutrient enrichment) than other secondary treatment methods. In areas of high nutrient levels, flagellated algae such as *Chlamydomonas* and *Euglena* (sometimes classified as a protozoan) often dominate. With lower nutrient levels, green algae such as *Chlorella*, *Scenedesmus*, and *Ankislrodesmus* will be the dominant types.

Protozoa show similar trends in oxidation ponds and other biological treatment processes. At the inlet where high organic loading is found, flagellated protozoa such as *Chilamonas* dominate. *Paramecium* and other free-swimming ciliates are the next dominant type to be found, and they feed on bacteria. As the bacterial population decreases, stalked ciliates such as *Vorticella* appear and feed on bacteria.

Larger organisms such as rotifers also use algae and help to produce an effluent low in BOD.

Although fungi have been found in oxidation ponds, they do not appear to be a significant component in the treatment process.

SLUDGES

ANAEROBIC DIGESTION. Anaerobic and facultatively anaerobic bacteria are the dominant organisms in anaerobically digesting sludges (Crowther and Harkness, 1975, and McKinney, 1962). Anaerobic organisms like fungi and protozoa normally are present in their inactive forms as spores and cysts. There are three stages that occur in the digestion process: *hydrolysis* (decomposition of a chemical compound by reaction with water), acid formation, and *methanogenesis* (formation of methane).

Large organic molecules present in sludge, such as fats, proteins, and polysaccharides, are split into smaller molecules by the insertion of water molecules in a process called *hydrolysis*. Bacteria responsible for this include *Clostridium, Micrococcus, Bacillus, Streptomycetes, Alcalegenes*, and *Pseudomonas*.

Many of the bacteria responsible for hydrolysis also are acid formers, and often hydrolysis and acid formation are considered as one step. The smaller organic molecules formed during hydrolysis are used by acid formers to produce carboxylic acids such as acetic and propanoic. *Aerobacter, Escherichia, Pseudomonas, Flavobacterium*, and *Alcaligenes* are examples of acid-forming bacteria.

Ultimately, the products of the acid-formation stage are converted to methane (CH_4) through methanogenesis. *Methanobacterium, Methanosarcina, Methanococcus*, and *Methanospirillum* have been identified in anaerobically digesting sludge.

The presence of *Desulfovibrio* in certain industrial sludges high in sulfates can convert sulfates to hydrogen sulfide, causing problems in the anaerobic digestion process.

AEROBIC DIGESTION. Little has been written on the organisms associated with aerobically digesting sludges. The bacterial types present are the same as in the activated sludge process, and the protozoa population also shows similar trends to that in activated sludge (Hartman *et al.*, 1979). Soon after digestion begins, free-swimming ciliates such as *Paramecium* and *Colpoda* dominate.

EFFLUENT

Many of the organisms found at the end of the secondary treatment process will find their way into the effluent before disinfection. With well-stabilized effluents, however, their populations should greatly be diminished by lack of food (McKinney and Gram, 1967).

To destroy organisms before discharge, the effluent may be disinfected (Hammer, 1975; McKinney, 1962; and Mitchell, 1974). While the complete destruction of all organisms or sterilization would ensure that *pathogens* (disease-causing organisms) would not be discharged, this is impractical and expensive. Disinfection, or the destruction of pathogens, is more practical. The success of the disinfection process is measured by the destruction of coliforms, a group of bacteria that act as indicator organisms. They are an acceptable measure of disinfection because the vegetative cells of most pathogenic organisms typically are more sensitive to disinfection than are the cells of nonpathogenic organisms. However, some pathogenic bacteria and protozoa may survive disinfection because their spore and cyst forms are more resistant than are the coliforms. *Enteric* (intestinal) viruses pose the most serious threat, as most disinfection methods in use today do not render them inactive.

The most common disinfection technique is oxidation, with chlorination the most widely used method of oxidation. Ozonation also has been used, especially in Europe and Canada, and offers a faster disinfection rate than chlorination. Other techniques, such as ultraviolet light or ionizing radiation, appear promising for future wastewater disinfection applications.

USE OF THE MANUAL

This text reviews the dominant microlife forms, autotrophic and heterotrophic bacteria (Chapter 4), and volatile acid and methanogenic bacteria (Chapter 5) whose life processes ultimately are responsible for the stabilization of wastewater. The text presents life processes or metabolic processes (Chapter 6) and those environmental factors, nutrients (Chapter 3), heavy metals (Chapter 7), and organics (Chapter 8) that affect the rate or success of these processes. Because the environmental factors that affect the life processes do so through their reaction with bacterial enzymes, a review of enzymes (Chapter 2) also is provided. Because enzymes and reaction or kinetic rates are involved in the life processes, the use of *bioaugmentation* (Chapter 9; a method of increasing the numbers of bacteria containing additional and desirable enzymes or favorable kinetic rates) is included. Operational or technical measures of monitoring the activity or health of the life processes also is provided (Chapters 10 and 11).

REFERENCES

Cooke, W.B. (1967) Trickling Filter Ecology. In *Biology of Water Pollution*. L.E. Keup *et al.*, (Eds.), U.S. Dep. of the Interior, Washington, D.C.

Crowther, R.F., and Harkness, N. (1975) Anaerobic Bacteria. In *Ecological Aspects of Used-Water Treatment*. Volume 1, C.R. Curds and H.A. Hawkes (Eds.), Academic Press, London, U.K.

Dhaliwal, B.S. (1979) *Nocardia amarae* and Activated Sludge Foaming. *J. Water Pollut. Control Fed.*, **51**, 344.

Gaudy, A.F., Jr., and Gaudy, E.T. (1980) *Microbiology for Environmental Scientists and Engineers*. McGraw-Hill, Inc., New York, N.Y.

Hammer, M.J. (1975) *Water and Wastewater Technology*. John Wiley & Sons, Inc., New York, N.Y.

Hartman, R.B., *et al.* (1979) Sludge Stabilization Through Aerobic Digestion. *J. Water Pollut. Control Fed.*, **51**, 2353.

Hawkes, H.A. (1983) Stabilization Ponds. In *Ecological Aspects of Used-Water Treatment*. Volume 2, C.R. Curds and H.A. Hawkes (Eds.), Academic Press, London, U.K.

McKinney, R.E. (1962) *Microbiology for Sanitary Engineers*. McGraw-Hill, Inc., New York, N.Y.

McKinney, R.E., and Gram, A. (1967) Protozoa and Activated Sludge. In *Biology of Water Pollution*. L.E. Keup, *et al.* (Eds.), U.S. Dep. of the Interior, Washington, D.C.

Mitchell, R. (1974) *Introduction to Environmental Microbiology*. Prentice Hall, Inc., Englewood Cliffs, N.J.

Pike, E.B., and Curds, C.R. (1971) The Microbial Ecology of the Activated Sludge Process. In *Microbial Aspects of Pollution*, G. Sykes and F.A. Skinner (Eds.), Academic Press, London, U.K.

Pipes, W.O. (1967) Bulking of Activated Sludge. In *Advances in Applied Microbiology*. Volume 9, W.W. Umbreit (Ed.), Academic Press, New York, N.Y.

Pipes, W.O. (1966) Ecological Study of Activated Sludge. In *Advances in Applied Microbiology*. Volume 8, W.W. Umbreit (Ed.), Academic Press, New York, N.Y.

Stowe, P.K. (1984) *The Georgia Operator*, **10**.

U.S. EPA (1977) *Aerobic Biological Wastewater Treatment Facilities*. Washington, D.C.

Water Pollution Control Federation (1990) *Wastewater Biology: The Microlife*. Special Publication, Alexandria, Va.

Chapter 2
Enzymes

In a wastewater treatment plant, wastes are broken down and used by large populations of organisms as sources of energy and building materials. These activities are facilitated by microbial enzymes, each of which can alter wastes in a specific way. An essential function of wastewater treatment facilities is to provide and maintain the conditions necessary for the proper functioning of microbes and their enzymes to best ensure efficiency in the degradation of organic wastes.

The emphasis of this chapter is on what enzymes are and the way they function. Included is a discussion of the effects of certain environmental variables on the abilities of microbes to degrade wastes. Microbial activities change with environmental conditions, influencing both the presence and activity of enzymes.

*C*HARACTERISTICS

Secondary treatment is the process by which organisms degrade organic wastes to water and simple gases. If degradation were complete, nothing would be left but inorganic salts, metals ions, and the microbes. However, microbial efficiency is limited by the fact that some organics are not able to be digested.

A great variety of organisms are present in wastewater, including different types of bacteria, protozoa, and rotifers. These organisms vary in the particular wastes they are capable of degrading, their methods of degradation, and in the products they form from the wastes. Some organisms consume the wastes directly, while others feed indirectly, consuming other organisms that have already ingested the wastes. Either way, almost every change in an ingested organic substance that an organism produces results from the action of a specific enzyme.

WHAT ARE ENZYMES?

Enzymes are protein molecules that regulate the rate of virtually every chemical reaction in living organisms. They are organic catalysts that increase the rate of these reactions without becoming changed themselves. Each chemical reaction catalyzed in an organism requires a unique enzyme; thus, enzymes are specific. Within a cell, enzymes are organized into sequential pathways, much like a production line in a factory.

WHAT DO ENZYMES DO?

Enzymes bind to reactant molecules (*substrates* or wastes) and then release them in a changed form (*products* or smaller substrates, water, and simple gases). In this manner, they do not cause chemical reaction to take place but rather increase their speed.

Enzymes can operate either by catalyzing the breakdown of a substrate to separate, smaller molecules or by joining substrate molecules and forming larger products. These substrates themselves may be carbohydrates, proteins, or lipids (fats). Because each substrate requires a unique enzyme, each enzyme catalyzes only one type of chemical reaction specific to it alone.

CLASSIFICATION BY CELLULAR LOCATION

Enzymes are divided into two large groups, based on whether they are used outside or inside a cell. Those used outside of a cell are secreted by the cell to the environment. Typically, these are digestive enzymes that degrade large, complex molecules to their smaller component parts. Complex carbohydrates such as starch are broken down to individual amino acids; fats and oils (lipids) are degraded to fatty acids and glycerol. Unlike when in the complex form, these smaller breakdown products are then able to enter cells.

Other enzymes are used within a cell. These can serve as carriers, transporting small molecules from outside to inside the cell, or they can be used to

---

transform small molecules from one type to another or link small molecules together to form larger ones. Enzymes within a cell can also be used to degrade small molecules to gain energy for growth and reproduction.

HOW NUTRIENTS GET INTO CELLS

Nutrients are chemicals required by a cell for it to live and reproduce. For these nutrients to get into cells, they must first be dissolved in the fluid surrounding the cells. Some nutrients can pass through a cell membrane by following a concentration gradient of higher nutrient concentration outside the cell to lower nutrient concentration inside the cell. Other nutrients are transported into the cell by carrier enzymes, some of which facilitate nutrient transport without requiring energy. Others rely on cellular energy for their action.

WHERE ENZYMES COME FROM

The living cells of every organism contain genetic blueprints for each enzyme the organism requires, with the proper "machinery" for their production. The blueprints are housed in materials called *genes*, which are made of deoxyribonucleic acid (DNA). Each gene contains the blueprint for one specific enzyme.

Some genes can be actively involved in enzyme production at any time under any condition. Others can be active only under specific environmental conditions. This means organisms are limited in their ability to adjust to environmental changes, depending on whether their genes are able to produce the required enzymes. For example, while some organisms are able to live in the presence of and even require oxygen (such as those in aeration basins), others may be killed by it (such as those in anaerobic digesters). Thus, environmental variables affect the life processes of organisms by influencing gene and, therefore, enzyme activity.

SENSITIVITY TO ENVIRONMENTAL CHANGE

Changes in environmental conditions can affect not only individual enzyme activity, but also the number and types of enzyme molecules produced by a cell. If environmental conditions are unfavorable to an organism, for example, fewer enzymes can be produced, and those that are may be of limited value.

Environmental factors that can influence enzymes include temperature, oxygen, acidity or alkalinity (pH), and the presence or absence of certain metals and salts. Thus, to obtain and maintain maximum efficiency from organisms in wastewater treatment systems, it is essential to closely monitor these variables.

Within wastewater treatment systems, the strength and composition of the waste determines the types of bacteria and, therefore, types of enzymes present. An example is that *Bacillus* bacteria have large numbers of enzymes that are particularly effective in digesting carbohydrates. Thus, *Bacillus* will tend to dominate in wastes from bakeries. *Pseudomonas* organisms are effective in digesting proteins and will tend to dominate in wastes from meat-packing plants.

Some wastewater treatment plants are now practicing *bioaugmentation*. This is a practice in which mixtures of microorganisms that produce unusually large amounts of digestive enzymes are seeded into a system so that the organic waste can be more effectively degraded.

SUMMARY

The secondary treatment process is a biological one fundamentally dependent on enzymes produced by organisms and used either outside or inside their cells. Enzyme synthesis is directed and regulated by genes within the cells. Enzymes secreted outside of cells break down complex molecules to simpler ones. The simple molecules (nutrients) then enter the cell by one of several processes, where they are converted by their other enzymes to cellular products or energy for growth and reproduction.

Organisms produce different enzymes under different environmental conditions. Therefore, successful management of the secondary treatment process involves manipulating the environment of organisms in a way that enables them to most efficiently degrade wastes.

SUGGESTED READINGS

Alberts, B., *et al.* (1983) *Molecular Biology of the Cell.* Garland Publishing, New York, N.Y.

Baker, J.J.W., and Allen, G.E. (1979) *A Course in Biology.* 3rd Ed., Addison Wesley, Reading, Mass.

Christensen, H.N. (1980) *Dissociation, Enzyme Kinetics, Bioenergetics.* 2nd Ed., Christensen, Ann Arbor, Mich.

Darnell, J., *et al.* (1986). *Molecular Cell Biology.* Sci. Am. Books, New York, N.Y.

Sheeler, P., and Bianchi, D.E. (1987) *Cells and Molecular Biology.* 3rd. Ed., John Wiley Publishers, New York, N.Y.

Stryer, L. (1981) *Biochemistry.* W.H. Freeman, San Francisco, Calif.

Chapter 3
Nutrients

The main objective of a biological wastewater treatment process is to remove pollutants such as suspended solids, biodegradable organic compounds, nitrogen, and phosphorus from wastewater. The efficiency of the process in removing pollutants depends on the interaction of three factors—organisms, pollutants, and environment. To remove the pollutants available as nutrients efficiently, organisms must have a growth environment with nutrients, temperature, pH, and mixing intensity in proper balance. The medium must have a carbon source, an energy source, an ample supply of inorganic nutrients and, in some cases, sources of organic nutrients.

Without an adequate growth environment and balanced nutrient conditions, the efficiency of the treatment process deteriorates rapidly from incomplete conversion of organic and inorganic substances to their products and inefficient removal of suspended solids. In that case, operators should be able to determine the conditions leading to the process failure.

In this chapter, the role of nutrients in microbial metabolism will be investigated. The sources of nutrients in wastewater, types of wastewater that may have nutrient deficiency, and the way an operator can solve a problem created by nutrient deficiency will be discussed. Biological removal of excessive nutrients from wastewater also will be addressed.

Every living cell contains nutrients, some of which are essential for cell growth. Other nutrients are used when present but are not essential. Carbon, oxygen, nitrogen, hydrogen, phosphorus, sulfur, potassium, calcium, magnesium, and iron are required in large amounts (Table 3.1). Sodium and chlorine are also required in large amounts by some, but not all, organisms.

Trace elements, which are used in small amounts by nearly all organisms, include cobalt, molybdenum, zinc, manganese, and copper. Other trace elements used by some organisms include selenium, nickel, tungsten, and boron. There also are specific organic compounds identified as growth factors that

Table 3.1   Elemental composition of microbial cells.

Elemental	Variations in composition of microbial cells, dry weight %
Carbon	45 – 55
Oxygen	16 – 22
Nitrogen	12 – 16
Hydrogen	7 – 10
Phosphorus	2 – 5
Sulfur	0.8 – 1.5
Potassium	0.8 – 1.5
Sodium	0.5 – 2.0
Calcium	0.4 – 0.7
Magnesium	0.4 – 0.7
Chlorine	0.4 – 0.7
Iron	0.1 – 0.4
All others*	0.2 – 0.5

*   Includes trace elements.

some bacteria are not capable of synthesizing for use as precursors or constituents of organic cell material. In such a case, they are introduced in chemical pollutants or may be released by metabolic activity of other organisms.

Although growth factor requirements vary widely from one organism to another, the major growth factors are amino acids, purines and pyrimidines, and vitamins.

COMPOSITION OF CELLS

ELEMENTAL COMPOSITION OF CELLS. Cells consist of 70 to 90% water and 10 to 30% dry material by weight (Neidhardt *et al.*, 1990). Of the dry material, 70 to 95% is organic and 5 to 30% inorganic. The inorganic content of the dry material forms the ash content of organisms once incinerated. The ash content of activated sludge from municipal treatment plants typically ranges between 20 and 35% (Pitter and Chudoba, 1990) but could be more than 50% depending on wastewater characteristics.

Carbon, oxygen, nitrogen, and hydrogen constitute approximately 92% of the dry mass of the bacterial cell. Although other elements are quantitatively less important, composing only 8% of the dry mass of the cell, they are important functionally. The composition of bacterial cells varies with the types of bacteria and growth conditions such as concentration and types of nutrients, temperature, pH, mixing intensity, and stage of growth. For example, *Acinetobacter* can store large amounts of phosphorus under certain conditions and can have a higher content of phosphorus than other organisms. Similarly, *Beggiatoa* and *Thiothrix* can store sulfur, resulting in a higher content of sulfur than other organisms. Fungal cells have a lower nitrogen content than bacterial cells and can grow in a medium containing lower nitrogen concentrations than those for bacterial cells.

The variations observed in elemental composition of bacterial cells are presented in Table 3.1. Carbon, for example, may constitute between 45 and 55% of the cell dry mass (Metcalf & Eddy, 1991). The term "All Others" in Table 3.1 includes the trace elements present in organisms. The content of trace and other elements can vary significantly depending on the concentration of these constituents in a growth medium. If the concentration of a trace element in the growth medium is higher than the minimum amount required for maximum growth, the element may be absorbed in excess of the organism's requirement. This results in a higher element content in the organism than is required. Nevertheless, the information on the trace element content of organisms is useful in determining their approximate requirements. Trace element content of organisms is given in Table 3.2 (Luria, 1960, and Pirt, 1975).

The composition of biomass developed in activated sludge processes, anaerobic digesters, and oxidation ponds has been reported by various investigators

Table 3.2   Trace element requirements of organisms.

	Dry weight of bacteria, %	
Element	**Luria (1960)**	**Pirt (1975)**
Manganese	0.002	0.005
Copper	0.008	0.001
Zinc	—	0.005
Cobalt	—	0.001
Molybdenum	—	0.001

Table 3.3   Composition of organic fraction of biomass (Metcalf & Eddy, 1991; Speece and McCarty, 1964; Oswald, 1963; and Varma and DiGiano, 1968).

	Dry weight of organic fraction, %				
	Activated sludge		**Anaerobic sludge**	**Algae**	
Element	**Mean**	**Range**		**Mean**	**Range**
Carbon	49.2	40.3 – 55.7	46.0	56.8	53.6 – 59.4
Oxygen	34.6	18.6 – 43.0	35.9	28.7	26.3 – 32.4
Nitrogen	8.8	4.9 – 16.3	11.1	8.6	7.0 – 9.6
Hydrogen	7.4	6.2 – 9.3	7.0	5.9	5.2 – 7.0

as summarized in Table 3.3 (Metcalf & Eddy, 1991; Speece and McCarty, 1964; Oswald, 1963; Varma and DiGiano, 1968). Because other elements are less important quantitatively, the organic fraction of biomass is assumed to consist of carbon, oxygen, nitrogen, and hydrogen. The organic fraction of activated sludge is, on the average, composed of 49.2% carbon, 34.6% oxygen, 8.8% nitrogen, and 7.4% hydrogen, though there are significant variations. Carbon, for example, varies from 40.3 to 55.7% of the organic fraction of activated sludges.

In general, activated sludges developed on wastes containing higher fractions of proteins have higher contents of nitrogen than those grown on wastes containing higher proportions of carbohydrates (Burkhead and Waddell, 1969). The nitrogen content of activated sludges also varies with the concentration of nitrogen in the growth medium. Symons and McKinney (1958) found that as the concentration of nitrogen in the growth medium decreased, the nitrogen content of activated sludge decreased from 9.7 to 1.5%. However, the ratio of carbon to hydrogen to oxygen remained essentially constant. While the type of substrate and the composition of the growth medium are found to affect the nitrogen content of activated sludge, the sludge age at 0.19 to 14.3 days influences neither the carbon content nor the nitrogen content of activated sludges (Weddle and Jenkins, 1971).

Speece and McCarty (1964) observed that solids retention time in the range of 5 to 30 days did not influence the composition of anaerobic biological solids grown on carbohydrates, fats, or proteins.

On the basis of the occurrence of four major elements in the composition of bacterial cells, the most commonly accepted empirical formula has been $C_5H_7O_2N$ (Hoover and Porges, 1952). When phosphorus is considered, one of the composition formulas given in Table 3.4 can be used. The nitrogen content of aerobic and anaerobic sludges varies between 8.9 and 12.2% of organic fraction of dry biomass, and phosphorus content between 1.2 and 2.3%.

Table 3.4   Empirical formulas for biomass.

	Organic fraction, %					
Empirical formula	Carbon	Oxygen	Hydrogen	Nitrogen	Phosphorus	References
Activated sludge						
$C_{60}H_{87}O_{23}N_{12}P$	52.4	26.8	6.3	12.2	2.3	McCarty (1970)
$C_{118}H_{170}O_{51}N_{17}P$	53	30.5	6.4	8.9	1.2	Sawyer (1956)
Anaerobic sludge						
$C_{54}H_{99}O_{32}N_{11}P$	44.9	35.4	6.9	10.7	2.1	Speece and McCarty (1964)
Algae						
$C_{106}H_{181}O_{45}N_{16}P$	52.4	29.7	7.4	9.2	1.3	Stumm and Tenney (1964)

The empirical biomass formulas are of value to treatment plant personnel, engineers, and scientists. They allow the calculation of nitrogen, phosphorus, and oxygen requirements. If compositions of biomass and wastewater are known, stoichiometric equations showing the proper proportion of elements can be written for the oxidation of substrate and synthesis of new cells, enabling the calculation of nitrogen, phosphorus, and oxygen requirements.

Formulas have been proposed for representing the organic composition of wastewater. Domestic wastewater can be represented by $C_{10}H_{19}O_3N$, lipid wastes by $C_8H_{16}O$, carbohydrate wastes by $C_6H_{12}O_6$, and proteinaceous wastes by $C_{16}H_{24}O_5N_4$ (Grady and Lim, 1980). Equation 3.1 can be written for the oxidation of domestic wastewater and the synthesis of new cells having a composition formula of $C_5H_7O_3N$. The observed yield coefficient has been assumed to be 0.23 kg (0.50 lb) volatile suspended solids formed/kg chemical oxygen demand removed.

$$C_{10}H_{19}O_3N + 3.625 \ O_2 + 0.775 \ NH_4^+ + 0.775 \ HCO_3^- \rightarrow$$
$$201(\text{mol wt}) \qquad\qquad 14(\text{mol wt})$$
$$1.775 \ C_5H_7O_2N + 1.9 \ CO_2 + 5.225 \ H_2O \quad (3.1)$$

On the basis of the stoichiometric relationship in Equation 3.1, 0.054 (0.775 × 14/201) lb nitrogen and 0.577 (3.625 × 32/201) lb oxygen are required for each pound of organic compounds removed from domestic wastewater (lb × 0.453 6 = kg). The phosphorus requirement can be calculated as one-fifth of the nitrogen requirement. However, if one of the composition formulas presented in Table 3.4 had been used, the phosphorus requirement could have been calculated from the stoichiometric relationship directly.

Elemental composition of biomass can also yield information about whether a nutrient deficit condition is occurring. Assume an analysis of organic fraction of dry biomass is conducted resulting in the following elemental composition: 52% carbon, 36% oxygen, 7% hydrogen, 4.5% nitrogen, and 0.5% phosphorus. When the results of this analysis are compared with those of a typical composition formula (Table 3.4), this biomass appears deficient in both nitrogen and phosphorus. Helmers *et al.* (1951) suggested a critical nitrogen level of 7% for the volatile fraction of biomass. If the nitrogen content of the volatile fraction of a biomass is below this level, nitrogen deficiency occurs. Studies by Helmers *et al.* (1951 and 1952) indicated that a volatile fraction of activated sludge contains between 0.74 and 2.25% phosphorus. A lower phosphorus level than 0.74% can indicate nutrient-deficient conditions.

The elemental composition of microbial cells varies with growth conditions. As a result, the nutrient requirements are expected to vary with growth conditions. Therefore, great care should be exercised in applying empirical formulas for determination of nutrient requirements.

BIOCHEMICAL COMPOSITION OF CELLS. Though nutrients within a bacterial cell also occur in small molecules such as water, organic substances, and as inorganic ions, they primarily occur in large molecules. Inorganic precursors such as ammonia (NH_3), orthophosphate (PO_4^{3-}), sulfide (HS^-), water (H_2O), and some other nutrients taken in by the cell are first synthesized to building blocks called *monomers*. Other nutrients have to be fixed or reduced to the same oxidation-reduction level as that of the element present in monomers before they can be used in biosynthesis. These nutrients include carbon dioxide (CO_2), nitrate (NO_3^-), nitrite (NO_2^-), nitrogen (N_2), sulfate (SO_4^{2-}), elemental sulfur (S), and thiosulfate ($S_2O_3^{2-}$). Thus, CO_2 has to be reduced to the same oxidation level as a carbohydrate or lower; NO_3^-, NO_2^-, and N_2 to amino groups ($-NH_2$); and SO_4^{2-}, S, and $S_2O_3^{2-}$ to sulfhydryl groups ($-SH$).

Large molecules (*polymers*) supplied as nutrients such as proteins, polysaccharides, and lipids have to be broken down to small organic and inorganic compounds that subsequently are synthesized to monomers. Still other nutrients have to be supplied as small organic molecules such as amino acids, purines and pyrimidines, and vitamins because they cannot be synthesized by some organisms. For example, *Streptococcus* and *Lactobacillus* require a variety of vitamins for biosynthesis of monomers. However, *Escherichia coli* can synthesize all of its macromolecules from a single carbon source.

Four classes of monomers are amino acids, simple sugars, fatty acids, and nucleotides, from which the cell eventually makes the large molecules (*macromolecules*) known as polymers. Macromolecules that make up the cell include proteins, polysaccharides, lipids, and nucleic acids (deoxyribonucleic acids [DNA] and ribonucleic acids [RNA]). Proteins are synthesized from amino acids, polysaccharides from simple sugars, and nucleic acids from nucleotides. Lipids are synthesized from compounds that include fatty acids, polyalcohols, simple sugars, amines, and amino acids (Stanier *et al.*, 1986). Macromolecules, their precursors, and elemental constituents of macromolecules are shown in Table 3.5.

Table 3.5   Macromolecules, building blocks, and constituents.

Macromolecules	Building blocks	Elemental constituents
Nucleic acids		C, H, O, P, N
RNA	Ribonucleotides	
DNA	Deoxyribonucleotides	
Proteins	Amino acids	C, H, N, O, S
Polysaccharides	Simple sugars	C, H, O
Lipids	Fatty acids, polyalcohols, simple sugars, amines, amino acids	C, H, O, N, P, S

Once the macromolecules are formed, they are assembled into higher order structures such as the cell wall, membranes, ribosomes, and chromosomes that are organized to form cells and organelles. An illustration of the formation of a cell is presented in Figure 3.1 (Becker, 1977).

The average macromolecular composition of cells can be estimated as 52.4% protein, 15.7% RNA, 3.2% DNA, and 28.7% other constituents (Gottschalk, 1986). However, wide variations in these values are observed with the types of organisms and growth conditions. For example, the protein content can vary between 50 and 100%; the DNA content can vary by two to three times; and the RNA, polysaccharide, and lipid contents can vary by 10 times or more.

In general, as the growth rate of organisms increases, the RNA content increases and the DNA content decreases, but the protein content remains

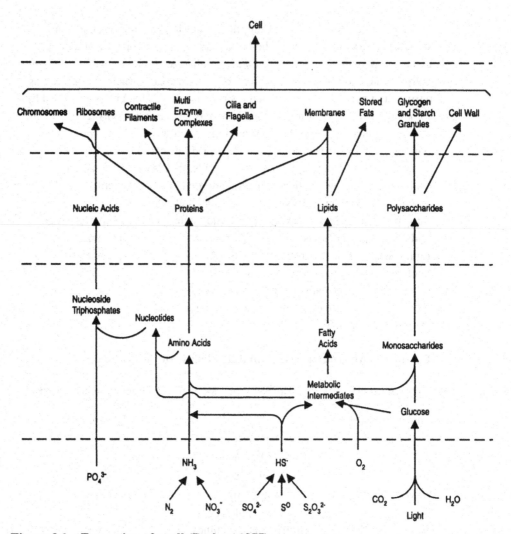

Figure 3.1 Formation of a cell (Becker, 1977).

relatively constant (Herbert, 1961, and Alroy and Tannenbaum, 1973). Herbert (1961) found that cell RNA and DNA contents primarily depended on the growth rate of the cell and were affected by the chemical composition of the growth medium only in so far as this affected the growth rate.

However, the case was different with polysaccharide and lipid synthesis. The cell polysaccharide and lipid contents were found to vary with the chemical composition of the growth medium. Herbert (1961) reported that the carbohydrate content of *Torula utilis* remained constant as the growth rate increased in a medium in which carbon was the growth-limiting nutrient. Under nitrogen limitation, however, the carbohydrate content of the cell decreased rapidly with increasing growth rate. The lipid synthesis showed a similar pattern as that of carbohydrate. The lipid content of *Bacillus megaterium* remained constant and low throughout the growth of the culture in a medium where carbon was limiting. When nitrogen was limiting, the lipid content of the cell increased rapidly as the growth rate was decreased.

Temperature has a significant effect on the macromolecular composition of cells. While the cell RNA content increases with a decrease in temperature, the cell DNA content remains approximately constant (Brown and Rose, 1969, and Tempest and Hunter, 1965).

Weddle and Jenkins (1971) reported that the lipid and protein nitrogen contents of activated sludge solids grown on primary wastewater did not vary significantly with changes in sludge age from 0.19 to 14.3 days. However, the carbohydrate content increased to approximately 24.5% from an average value of 12.7% as the sludge age was decreased from 1 to 14.3 days to 0.2 to 0.25 days, but it remained essentially constant in the practical range of operation of plants. Genetelli (1967) observed that the DNA content of activated sludge was a function of quantity and types of organisms present. The DNA content was observed to decrease as the sludge proceeded from being a zoogleal sludge to a more filamentous mass presented by *Sphaerotilus*. Studies by Rickard and Gaudy (1968) indicated that the RNA, DNA, and protein contents of activated sludge remained relatively constant, but the carbohydrate content decreased with an increase in the mixing intensity (mean velocity gradient).

*F*ORMS AND FUNCTIONS OF NUTRIENTS IN BACTERIAL METABOLISM

Compounds are taken up by cells from the environment to carry out primarily two basic metabolic activities: energy production (*bioenergetics*) and synthesis of new cell material (*biosynthesis*). Organisms produce energy from light, inorganic and organic compounds. Inorganic compounds that serve as the source of energy include ammonium (NH_4^+), nitrite (NO_2^-), dissolved sulfide (H_2S), elemental sulfur (S), thiosulfate ($S_2O_3^{2-}$), and ferrous iron (Fe^{2+}). These compounds are oxidized and the energy released during oxidation is used for cell maintenance, for synthesis of new cell material, and for cell movement if the organism is *motile*. Some compounds also serve as nutrients for the synthesis of new cell material. Forms of nutrients and their functions in bacterial metabolism are presented in Table 3.6.

CARBON. Carbon is the main element used during biosynthesis. It makes up approximately 50% of the bacterial dry mass and is found in all macromolecules of the cell. Its source is either organic compounds such as amino acids, fatty acids, organic acids, sugars, nitrogenous bases, aromatic compounds, and many other organic substances, or carbon dioxide. During the

Table 3.6   Sources and functions of major elements in bacteria (Gottschalk, 1986).

Element	Sources	Function in metabolism
Carbon	Organic compounds, CO_2	Main constituent of cellular material
Oxygen	O_2, H_2O, organic compounds, CO_2	Main constituent of cellular material
Hydrogen	H_2, H_2O, organic compounds	Main constituent of cellular material
Nitrogen	NH_4^+, NO_3^-, N_2, organic compounds	Main constituent of cellular material
Sulfur	SO_4^{2-}, HS^-, S, $S_2O_3^{2-}$, organic sulfur compounds	Constituent of some amino acids and some vitamins
Phosphorus	HPO_4^{2-}	Constituent of nucleic acids, phospholipids, and nucleotides
Potassium	K^+	Principal inorganic ion in the cell, cofactor of some enzymes
Magnesium	Mg^{2+}	Cofactor of many enzymes present in cell walls and membranes
Calcium	Ca^{2+}	Cofactor of many enzymes present in cell walls and endospores
Iron	Fe^{2+}, Fe^{3+}, Organic iron complexes	Present in cytochromes and ferrodoxins, cofactor of enzymes
Sodium	Na^+	Involved in various transport processes provides cell wall stability, and enzyme activity
Chlorine	Cl^-	Important inorganic anion in the cell

synthesis of cellular carbon from organic compounds, a portion of carbon is excreted from the cell, either as carbon dioxide or as organic waste products. The carbon dioxide is, in turn, used by autotrophic and other bacteria as the primary source of cellular carbon. It is obtained from the aqueous environment surrounding the bacterial cell, not from the air.

OXYGEN AND HYDROGEN. Oxygen and hydrogen are the main constituents in the cellular material. The source of the cellular oxygen can be molecular oxygen, water, organic compounds, or carbon dioxide, and the source of the cellular hydrogen can be molecular hydrogen, water, and organic compounds. Oxygen, as an electron acceptor, is used to classify organisms. Organisms that use oxygen are called *aerobes* and those that do not are *anaerobes*. *Obligate aerobes* must have oxygen for growth, while *obligate anaerobes* cannot grow in its presence and sometimes are killed by oxygen.

Aerotolerant anaerobes can tolerate low levels of oxygen without any harm. *Microaerophilic* organisms are aerobes that grow better at reduced levels of oxygen, while *facultative* organisms can grow both in the presence and absence of oxygen.

NITROGEN. Nitrogen comprises 14% of the cellular material and is the major constituent of proteins and nucleic acids of the cell. Nitrogen also is found in *peptidoglycan*, the rigid cell wall layer of most bacteria. Nitrogen occurs in nature in both organic and inorganic forms. Organic forms include amino acids and nitrogenous bases, which are the breakdown and mineralization products of dead organisms. Organic nitrogen in wastewater is partially available for nutritional use. Studies by Helmers *et al.* (1952) indicate that 64 to 80% of organic nitrogen in domestic wastewater is available to organisms.

Inorganic nitrogen that can be used by organisms typically is in the form of nitrogen gas (N_2), ammonia-nitrogen ($NH_3 + NH_4^+$), nitrite (NO_2^-), and nitrate (NO_3^-). Nitrogen gas from the atmosphere is available to the nitrogen-fixing bacteria only, such as *Azotobacter*, *Bacillus*, cyanobacteria, *Clostridium*, and *Rhodobacter*. In the fixing process, nitrogen gas first is reduced to ammonium (NH_4^+) and then is converted to organic nitrogen. Ammonia-nitrogen is found in wastewater as ammonia, which can be either in the form of unionized gaseous nitrogen or the ammonium ion:

$$NH_3 + H_2O \leftrightarrows NH_4^+ + OH^- \tag{3.2}$$

Relative concentrations of ammonia and ammonium ion depend on the pH of wastewater. Ammonium ion is the principal form below pH 9, and ammonia the principal form above pH 9.7 (Sawyer and McCarty, 1978).

Ammonium ion is the most readily usable form among all inorganic forms of nitrogen. Its use does not require an oxidation-reduction reaction because its nitrogen atom is at the same oxidation level as the nitrogen atom of amino acids and purines and pyrimidines. Amino acids are precursors for proteins, and purines and pyrimidines are precursors for nucleotides. Ammonium ions can be assimilated by organisms through three reactions, called *aminations*, in which it forms the amino group of glutamic acid and the amino groups of asparagine and glutamine. These groups, except those in asparagine, are transferred to form all other nitrogenous precursors of cellular macromolecules. Ammonia-nitrogen can be considered as 100% available for nutritional use by bacteria.

Many bacteria also use nitrite and nitrate as the sole nitrogen sources. However, only 30 to 70% of nitrite-nitrogen or nitrate-nitrogen is available for nutritional use because nitrate (or nitrite) has to be reduced from a $^{+5}$ oxidation state ($^{+3}$ for nitrite) to the oxidation level of ammonia, which is $^{-3}$ for use as a nitrogen source for growth (Gerardi, 1991).

This process of reduction, known as assimilative nitrate reduction, requires energy. Organisms that obtain energy from other sources, such as oxidation of organic compounds, have to use a greater portion of that energy when nitrate (or nitrite) is available as the sole source of nitrogen compared to when ammonia nitrogen is available. This results in lower biomass yield than the biomass yield obtained with ammonia-nitrogen. Most photosynthetic and some nonphotosynthetic organisms can assimilate nitrite or nitrate. However, both oxidized nitrogen forms must eventually be reduced to ammonia before they can be incorporated to amino acids (Grady and Lim, 1980).

PHOSPHORUS. Phosphorus constitutes approximately 3% of the cell dry weight. Cellular components containing phosphorus are high-energy compounds such as adenosine triphosphate (ATP) and adenosine diphosphate (ADP), nucleotides such as nicotinamide adenine dinucleotide (NAD) and flavin adenine dinucleotide (FAD), nucleic acids, and phospholipids. Phosphorus is involved with virtually all facets of metabolism, such as the biosynthesis of proteins, nucleic acids, complex carbohydrates, lipids, and other cellular constituents (Hodson, 1973).

Phosphorus occurs in nature in the form of phosphate salts such as calcium, iron, and aluminum phosphates and their organic and inorganic derivatives. Inorganic phosphates consist of orthophosphates, which are soluble, and polyphosphates and metaphosphates, which are both relatively insoluble. Metaphosphates are rare in nature and commerce. Polyphospates *hydrolyze* (combine with water) in aqueous solutions to yield orthophosphates.

Organisms play a significant role in the hydrolysis of polyphosphates. Their acidic metabolic products such as organic, nitric, and sulfuric acids solubilize calcium phosphate, and their production of hydrogen sulfide results in the dissolution of ferric phosphates. Most organisms incorporate orthophosphates to their cellular material. Most organic phosphates, however, cannot be used before they are broken down, releasing free inorganic phosphate by the action of enzymes called phosphatases.

SULFUR. Sulfur is essential for the synthesis of proteins. Cysteine and methionine, 2 of the 20 common amino acids (monomers of proteins), contain sulfur atoms. Sulfur also is present in a number of vitamins that are used as growth factors. Sulfur occurs in nature in many forms, though only four account for most sulfur in nature: sulfhydryl (R-SH), sulfide (HS^-), elemental sulfur (S), and sulfate, SO_4^{2-}. All these forms can be used by bacteria.

Sulfur occurs almost exclusively in reduced forms as -SH or -S-S- groups in living organisms (Stanier *et al.*, 1986). Compounds that can be incorporated directly to the organic molecules of the cell include sulfide (HS^-) and sulfur-containing amino acids and vitamins. The other sulfur compounds (sulfate, elemental sulfur, and thiosulfate) have to be reduced to the same oxidation level as sulfide for them to be incorporated to cell material. Reduction of sulfur atoms in these compounds is carried out by a number of organisms, including higher plants, algae, fungi, and many bacteria.

Once the sulfide is formed, it is incorporated to cell material through a reaction with the amino acid serine to form the amino acid cysteine. Other sulfur-containing organic compounds are then synthesized from the sulfur of cysteine. When organic compounds are available as sources of sulfur they are mineralized, liberating sulfide, sulfur-containing amino acids and vitamins, which can be incorporated directly to the cell material. Some bacteria cannot reduce sulfur compounds and depend on the availability of reduced sulfur compounds. For example, methanogenic bacteria require hydrogen sulfide as a sulfur source for growth (Gottschalk, 1986).

OTHER NUTRIENTS. While carbon, oxygen, hydrogen, nitrogen, phosphorus, and sulfur are needed for the synthesis of macromolecules of the cell, other elements are required for three basic functions: (1) as enzyme activators called either coenzymes or metal cofactors, (2) to transfer electrons in oxidation-reduction reactions, and (3) to serve as regulators of osmotic pressure.

Potassium is required in activating enzymes, including those involved in protein synthesis, and in the maintenance of osmotic pressure and regulation of pH. Magnesium activates many enzymes, including those involved in phosphate transfer, and is used in binding enzymes to substrates. It is a constituent of chlorophylls and present in cell walls, membranes, ribosomes, and phosphate esters.

Calcium is a cofactor for some enzymes such as proteinases. It is present in cell walls, where it provides structural integrity. It is found in bacterial endospores as calcium-dipocolinate, which increases the heat resistance of spores. Sodium may be required by certain organisms, especially those growing in marine and salty environments and by methane-producing bacteria.

Chlorine is an important *anion* (a negatively charged ion) in the cell, and iron is essential because it is found in a number of enzymes, including certain electron-transporting enzymes of respiration such as iron-sulfur proteins and cytochromes. Iron occurs in nature as ferric (Fe^{3+}) and ferrous (Fe^{2+}) forms and organic iron compounds. Ferric forms are found under aerobic conditions and reduced ferrous forms are found under anaerobic conditions.

Cobalt is required for the formation of vitamin B_{12}, and zinc is essential for the activity of many enzymes such as carbonic anhydrase, alcohol dehydrogenase, and RNA and DNA polymerases. It holds the subunit of the enzymes in proper configuration so that the enzymes can be active. Molybdenum is found in certain enzymes involved in assimilatory nitrate reduction and N_2 reduction. Copper plays a similar role as that of iron in respiration. It is present in cytochrome oxidase, in nitrite reductase of denitrifying bacteria, and in oxygenases.

Manganese acts as an activator in certain enzymes and also is found in some enzymes that detoxify toxic forms of oxygen. Nickel is a constituent of enzymes known as hydrogenases, which are involved in H_2 uptake or evolvement. Tungsten and selenium are a part of the enzyme system involved in metabolizing formate.

AMINO ACIDS, PURINES AND PYRIMIDINES, AND VITAMINS.
Amino acids, purines and pyrimidines, and vitamins are provided as nutrients for organisms incapable of synthesizing them. While of these compounds are

used as precursors for building blocks, others are used as constituents of organic cell material.

Vitamins are organic compounds that serve as neither energy sources nor building blocks of macromolecules but function as parts of enzymes—either coenzymes or a part of coenzymes or prosthetic groups—that are catalysts for biological reactions (see Chapter 6). Vitamins most commonly required by organisms include thiamine (vitamin B_1), biotin, pyridoxine (vitamin B_6), and cobolamin (vitamin B_{12}) (Brock and Madigan, 1988).

NUTRIENT TRANSPORT

Once nutrients are available to the cell they are transported across the cell membrane to the inside of the cell. However, proteins, polysaccharides, nucleic acids, lipids, and other large molecules have to be broken down before they can be transported. Their breakdown is accomplished by the action of extracellular (functioning outside the cell) enzymes called *exoenzymes*, which are produced in the cell and secreted through the cell barrier to act on large molecules.

Most bacteria produce exoenzymes to break down proteins to amino acids, polysaccharides to simple sugars, nucleic acids to purines and pyrimidines, and lipids to fatty acids. Along with inorganic ions, these subunits are then conveyed inside the cell by *passive* and *active processes* as shown in Figure 3.2.

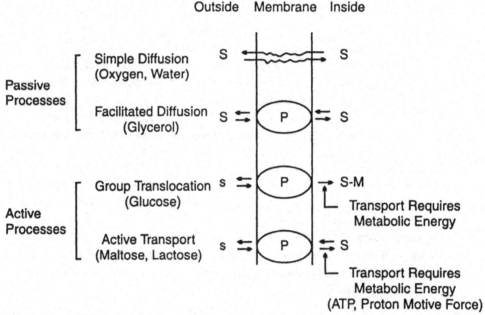

Figure 3.2   Schematic of various classes of membrane transport: S and s = high and low concentration, respectively, of nutrients; P= carrier protein; S-M = chemically altered nutrient (Stanier *et al.*, 1986).

PASSIVE TRANSPORT. In passive processes, the cell does not spend any energy to transport the nutrients across the cell membrane. The passive processes are *simple* and *facilitated diffusion.*

Simple Diffusion. Simple diffusion occurs across a semipermeable cell membrane in the direction from an area of high concentration to an area of low concentration. Small molecules—such as water, oxygen, and carbon dioxide—and some small nonpolar and fat soluble substances—such as fatty acids and alcohols—may pass across the cell membrane by simple diffusion. At equilibrium, the concentration of the nutrient outside and inside the cell is equal (Figure 3.2).

Facilitated Diffusion. The facilitated diffusion process is mediated by specific membrane proteins known as *permeases* or *carrier proteins.*

CARRIER PROTEINS. A carrier protein binds to a specific nutrient on the outer surface of the cell membrane. The complex formed between the nutrient and carrier protein diffuses across the membrane and dissociates inside the cell, releasing the nutrient. Facilitated diffusion occurs in the direction from a higher to a lower nutrient concentration. At equilibrium, the concentration of the nutrient inside and outside the cell is equal.

Concentration of nutrients inside the cell may be higher than those in the environment. Therefore, passive processes may not be effective in providing adequate nutrient supply for cell growth. In such cases, the cell uses energy to transport the nutrients inside the cell by the active processes of *group translocation* and *active transport.* The energy used during the nutrient transport can be derived from the breakdown of organic or inorganic compounds or from light.

GROUP TRANSLOCATION. In group translocation, the nutrient is chemically altered during the transport process. Once the chemically altered nutrient is inside the cell, it is trapped there as the cell membrane is impermeable to this new form of the nutrient. As a result, high levels of the chemically altered nutrient can be concentrated inside the cell. Substances such as glucose, mannose, fructose, and possibly purines, pyrimidines, and fatty acids are transported by group translocation in some bacteria (Neidhardt *et al.*, 1990).

ACTIVE TRANSPORT. In the process of active transport, a carrier protein of the cell membrane combines with a nutrient. Once inside the cell, the nutrient is released without chemical alteration. Thus, if not consumed in cell reactions, the concentration of the nutrient inside the cell can be several hundred to a thousand times greater than that outside the cell. The active transport process requires energy to transport the nutrient against the concentration gradient. The energy obtained from the breakdown of organic or inorganic compounds or from light is used to establish a proton gradient across the cell membrane or used in the form of high-energy compounds, especially ATP.

In bacteria, the protons are hydrogen ions, with a concentration higher outside than inside the cell. The electrochemical potential of this proton separation is used to pump nutrients from outside the cell to inside. The energy obtained from the ATP also is used to transport the molecules.

Substances that are actively transported include some sugars; amino acids; organic acids; and a number of inorganic ions such as sulfate, phosphate, and potassium. While some nutrients are transported by group translocation in some bacteria, the same nutrients may be transported by active transport in other bacteria. For example, glucose is transported by the mechanism of group translocation in *E. coli*, *Bacillus subtilis*, *Clostridium pasteurianum* and by active transport in *Pseudomonas aeruginosa*, *Micrococcus luteus*, and *Mycobacterium smegmatis*.

SOURCES OF NUTRIENTS FOUND IN WASTEWATER

Sources of nutrients in wastewater include domestic, commercial, industrial, and agricultural wastes; domestic water supply; surface runoff; and sewer infiltration. Carbon, hydrogen, and oxygen occur in a wide variety of organic constituents of domestic, commercial, industrial, and agricultural wastes such as proteins, carbohydrates, fats, oils, and grease. They are also found in inorganic constituents of wastes. Carbon is available from the components of alkalinity—mainly bicarbonates, carbonic acid, and carbonates. Oxygen is available from dissolved oxygen, water, and bicarbonate, and hydrogen from hydrogen gas and water.

NITROGEN COMPOUNDS. Nitrogenous compounds occur in wastewater as organic and inorganic nitrogen. Organic nitrogen does not include all organic nitrogen compounds, but organically bound nitrogen in the trinegative oxidation state (APHA, 1989). Inorganic nitrogen consists of ammonia (NH_3), nitrite (NO_2), and nitrate (NO_3). Sources of nitrogen in wastewater are primarily domestic, industrial, commercial, and agricultural wastes.

Organic nitrogen concentration in untreated domestic wastewater typically is in the range of 8 to 35 mg/L as nitrogen. *Urea*, one of the major sources of ammonia, is an organic nitrogen compound and is rapidly *hydrolyzed* (split with the addition of water) to ammonia and carbon dioxide by *Proteus mirabilis* and *P. vulgaris*. Ammonia-nitrogen concentrations typically range between 12 and 50 mg/L in untreated wastewaters. However, concentrations of both nitrite-nitrogen and nitrate-nitrogen are negligible, normally constituting only 1% of total nitrogen (U.S. EPA, 1975).

The nitrogen content of industrial wastewater varies. Industries that tend to have a high content of nitrogen in their wastewater include meat-processing plants, milk-processing plants, petroleum refineries, ice plants, fertilizer manufacturers, certain synthetic fiber plants, and industries using ammonia for scouring and cleaning operations.

Feedlot runoff also constitutes a major source of nitrogen. Ammonia-nitrogen concentrations as high as 300 mg/L and organic nitrogen concentrations of up to 600 mg/L have been reported (U.S. EPA, 1975).

Urban runoff can be a significant source of nitrogen in wastewater. Concentrations of total nitrogen in urban runoff have been found to vary between 2.1 and 2.7 mg/L. The organic nitrogen content, though, has been reported to be 0.85 mg/L.

The runoff from agricultural land treated with artificial fertilizers may contain significant quantities of nitrogen. Runoff from various types of agricultural lands has been found to contain 1.2 to 9.0 mg/L total nitrogen.

Nitrate concentrations in surface waters typically are low, but can attain high levels in some groundwater. Snoeyink and Jenkins (1980) reported nitrate concentrations of 0.41, 0.10, and 13 mg/L for reservoir water, river water, and well water, respectively. Treated wastewater from nitrifying biological treatment plants can contain up to 30 mg/L nitrate-nitrogen (APHA, 1989).

PHOSPHORUS COMPOUNDS. Phosphorus compounds found in wastewater include orthophosphates, condensed phosphates, and organic phosphorus. Orthophosphates (PO_4^{3-}, HPO_4^{2-}, $H_2PO_4^-$, and H_3PO_4) primarily are represented by HPO_4^{2-} at the neutral pH levels observed in domestic wastewater. The forms $H_2PO_4^-$ and H_3PO_4 predominate in acidic wastewaters, and PO_4^{3-} in alkaline wastewater. Phosphorus in the orthophosphate form is available to bacteria for metabolism without further breakdown.

Condensed phosphates include compounds with two or more orthophosphate groups linked together with the characteristic P-O-P linkage. Condensed phosphates consist of *polyphosphates*, which are linear molecules, and *metaphosphates*, which are cyclic (Snoeyink and Jenkins, 1980). Condensed phosphates undergo hydrolysis in wastewater and are converted to the orthophosphate forms, though the hydrolysis process is slow.

Organic phosphorus compounds also are converted to orthophosphates by hydrolysis and occasionally by oxidation or *isomerization*—the process of changing the structure of a compound but not the molecular formula. The concentration of organic phosphorus compounds in domestic wastewater typically is low but can be significant in industrial wastes and wastewater sludges.

Snoeyink and Jenkins (1980) reported that organic phosphates in fresh, raw domestic wastewater were less than 1 mg/L as phosphorus. The distribution of other phosphates as phosphorus was as follows: orthophosphate—5 mg/L, tripolyphosphate—3 mg/L, and pyrophosphate—1 mg/L. Typical concentrations of total phosphorus were reported to range between 3 and 15 mg/L as phosphorus.

The major sources of phosphorus in wastewater are human wastes and detergents. Human wastes account for about 30 to 50% of the phosphorus in domestic wastewater as a result of breakdown of proteins and nucleic acids and liberation of phosphorus in urine. Polyphosphates are used as "builders" in detergents and can contribute as much as 50 to 70% of total phosphorus content of domestic wastewater. Because phosphorus is one of the nutrients responsible for deterioration of water quality in receiving waters, a number of states have passed laws banning the use of phosphates in detergents. As a result, significant reductions in phosphorus content of wastewater have been reported.

Chemicals containing polyphosphates sometimes are used to control corrosion and scale formation in water supply systems. The contribution from domestic water supply can range between 2 and 20% of the phosphorus in domestic wastewater. Certain industries can also add significant quantities of phosphorus to domestic wastewater, including potato and flour processing, fertilizer manufacturing, metal-finishing industries, dairies, commercial laundries, slaughterhouses, and animal feedlots (U.S. EPA, 1976).

SULFUR COMPOUNDS. Sulfur is found in domestic, commercial, and industrial wastes and in the domestic water supply in the forms of sulfate, sulfides, sulfites, and organic sulfur. The sulfate concentration in untreated domestic wastewater typically ranges from 20 to 50 mg/L, while domestic water supply can have sulfate concentrations as high as 250 mg/L (Sawyer and McCarty, 1978). Therefore, sulfate concentrations in wastewater can vary from a few to hundreds of milligrams per liter.

Sulfides occur in wastewater because of the decomposition of organic matter, mostly the bacterial reduction of sulfate. They are found in the form of dissolved sulfides (H_2S, HS^-, S^{2-}) and insoluble metallic sulfides. The sulfide ion S^{2-} is essentially absent over the pH range typically observed in domestic wastewaters. Therefore, the principal forms of dissolved sulfides in domestic wastewaters are H_2S and HS^-. Concentrations of H_2S in wastewater range between 3 and 4 mg/L (U.S. EPA, 1985). Insoluble metallic sulfides include pyrite (FeS_2), smythite (Fe_3S_4), and pyrrhotite, which varies in composition from FeS to Fe_4S_5 (U.S. EPA, 1974). Sulfite ions (SO_3^{2-}) can occur in industrial waste discharges and treatment plant effluents subjected to dechlorination with sulfur dioxide (SO_2). Organic sulfur compounds found in wastewater are principally in three forms: mercaptants, thioesters, and disulfides. The concentration of organic sulfur compounds ranges between 1 and 3 mg/L in domestic wastewater (U.S. EPA, 1985).

OTHER NUTRIENTS. The sources of other nutrients in wastewater—potassium, magnesium, sodium, calcium, iron, manganese, zinc, chloride, copper, nickel, selenium, cobalt, and molybdenum—include the domestic water supply, water supply chemicals, industrial wastes, irrigation return wastes, surface runoff, sewer infiltration, and groundwater. Groundwaters from soil and rock formations can contain higher concentrations of some nutrients than do surface waters (Wood and Tchobanoglous, 1975). For example, groundwater passing through a limestone formation may have higher concentrations of calcium, magnesium, sulfate, and chloride. Groundwater can also contain higher concentrations of iron and manganese in locations where anaerobic conditions exist under which insoluble ferric forms are reduced to soluble ferrous ions.

Domestic and industrial use of the water supply can add significant quantities of these nutrients to wastewater. Surface runoff and sewer infiltration also can be important sources of these nutrients in wastewater and may be washed from the surface and find their way to the sewer system through manhole covers and sewer interconnections. Infiltration caused by damaged or improperly laid sewers leads to additional dissolved minor nutrients in wastewater.

NUTRIENT REQUIREMENTS IN BIOLOGICAL TREATMENT OF WASTEWATER

To achieve efficient removal of nutrients in a biological treatment process, they must be available in adequate amounts. If any essential nutrients are absent, microbial growth will not occur.

Early investigators reported that nutrients such as organic carbon (expressed as 5-day biochemical oxygen demand [BOD_5]), nitrogen (N), and phosphorus (P) were required in the ratio of 100:5:1 for efficient treatment of wastewater. However, later studies indicated that nutrient requirements are not constant, but vary from plant to plant depending on the type of processes applied (aerobic or anaerobic), composition of waste, plant design, mode of operation, and environmental factors (Broderick and Sherrard, 1985, and Sherrard and Schroeder, 1976).

MICROBIAL GROWTH EQUATION. Several methodologies exist to determine the nutrient requirements of organisms. One technique allows estimation of nitrogen, phosphorus, and oxygen requirements by writing an equation for microbial growth. This technique requires knowledge about the composition of wastewater and organisms, kinetic constants, and the sludge age at which the plant is operated. Kinetic constants and the sludge age are used to determine the amount of cells (volatile suspended solids [VSS]) produced per unit weight of substrate removed using Equation 3.3:

$$Y_{obs} = \frac{Y_{max}}{1 + k_d \theta_c} \qquad (3.3)$$

Where

Y_{obs} = the observed yield coefficient, lb cells (or VSS) produced/lb BOD5 removed (kg/kg);

Y_{max} = the maximum yield coefficient, lb cells (or VSS) produced/lb BOD_5 removed (kg/kg);

k_d = the decay coefficient, day^{-1};

θ_c = the sludge age, lb solids in aeration tanks/lb solids wasted per day (kg/kg · d).

The observed yield is used to write the equation for microbial growth from which nitrogen, phosphorus, and oxygen requirements can be determined, as illustrated in Equation 3.1 for aerobic processes. Assuming a wastewater composition of $C_6H_{12}O_6$ and a cell composition of $C_{60}H_{87}O_{23}N_{12}P$, nitrogen and phosphorus requirements for activated sludge processes have been calculated and converted to BOD_5:N:P, or COD:N:P ratios. The variation of BOD_5:N:P or COD:N:P ratios with sludge age, maximum yield coefficient, and decay coefficient are shown in Figures 3.3 and 3.4. The variation of oxygen requirement with sludge age is presented in Figure 3.5. Figure 3.3 can be used to determine the nitrogen and phosphorus requirements.

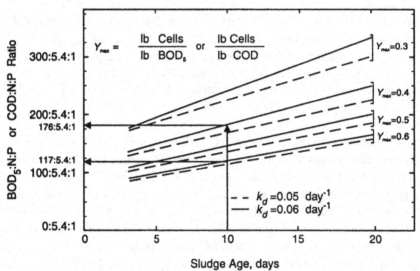

Figure 3.3 Variation of nutrient requirements with sludge age and yield and decay coefficients in activated sludge processes (lb × 0.453 6 = kg).

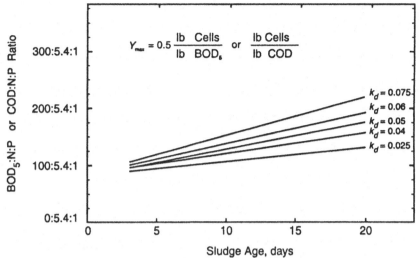

Figure 3.4   Effect of decay coefficient on nutrient requirements at different sludge ages in activated sludge processes (lb × 0.453 6 = kg).

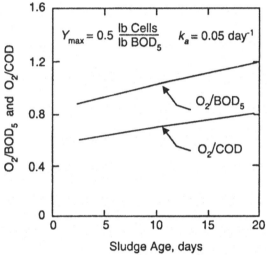

Figure 3.5   Variation of oxygen-to-substrate ratio with sludge age in activated sludge processes (lb × 0.453 6 = kg) (Sherrard and Schroeder, 1976).

Example 1—calculate the BOD_5:N:P and COD:N:P ratios for an activated sludge process.

Given:
- Sludge age, θ_c = 10 days;
- Maximum yield coefficient, Y_{max} = 0.6 lb VSS produced/lb BOD_5 removed (kg/kg);
- Decay coefficient, k_d = 0.06 day^{-1}; and
- BOD_5:COD ratio of wastewater = 0.67.

Solution:

Step 1
- Read BOD_5:N:P ratio from Figure 3.3 directly for θ_c = 10 days and Y_{max} = 0.6. The BOD_5:N:P ratio is 117:5.4:1.

Step 2
- Express Y_{max} in terms of COD. Y_{max} = 0.6 lb VSS/lb BOD_5 × 0.67 = 0.4 lb VSS/lb COD.
- Read COD:N:P ratio from Figure 3.3 directly for θ_c = 10 days and Y_{max} = 0.4. The COD:N:P ratio is 176:5.4:1.

The ratio of BOD_5:N:P or COD:N:P increases as sludge age increases. Nitrogen and phosphorus requirements decrease as sludge age increases because fewer cells (or sludge) are produced at higher sludge ages. Sherrard and Schroeder (1976) found that the BOD_5:N:P ratio of 100:5:1, previously suggested for efficient treatment, corresponded to the nutrient requirements at low sludge age operation, such as 3 days.

Although maximum yield coefficient is assumed constant in determination of nutrient requirements, it is not constant, but varies with the nature of wastewater, types of organisms, process type (aerobic or anaerobic), pH, and temperature. However, it typically lies between 0.2 and 0.3 kg (0.4 and 0.6 lb) VSS produced/kg (lb) COD consumed for aerobic heterotrophs. As the maximum yield coefficient increases, the nutrient requirements also increase. The reason for this is that the higher the maximum yield coefficient, the higher the cell production and, consequently, the higher the nitrogen and phosphorus requirements will be. For a given waste strength (BOD_5, COD), the maximum yield coefficient obtained from carbohydrate wastes typically is higher than that from proteinaceous wastes (Wood and Tchobanoglous, 1975). For this reason, carbohydrate wastes require higher quantities of other nutrients.

Nutrient requirements also vary with decay coefficients. As the decay coefficient increases, nitrogen and phosphorus requirements decrease, as shown in Figure 3.4. At higher decay coefficients, less cells are produced, resulting in lower nitrogen and phosphorus requirements, thus higher BOD_5:N:P or COD:N:P ratios.

Oxygen requirement increases along with sludge age. At higher sludge ages, a higher portion of organic substrate is oxidized to carbon dioxide and nitrate, with a smaller portion used for cellular synthesis. The variation of oxygen:BOD_5 and oxygen:COD ratio with sludge age is shown in Figure 3.5 for activated sludge processes.

Anaerobic processes typically are used to stabilize domestic wastewater sludges and waste activated sludges, both of which are assumed to contain adequate amounts of nutrients to support anaerobic stabilization. However, concerning the anaerobic treatment of nutrient deficient, dilute industrial wastewater, nutrient addition may be necessary. As shown in Table 3.4, the nitrogen and phosphorus contents of anaerobic and aerobic organisms are similar, thus the nutrient requirements of both types of organisms (amount of nitrogen and phosphorus required per unit weight of organisms produced—kg N or kg P required/kg VSS produced) should be the same.

However, because the number of new cells produced in anaerobic processes is less than in aerobic processes, less nitrogen and phosphorus are required in the new cell material. In short, anaerobic processes require substantially less nitrogen and phosphorus than aerobic processes.

Typical values of yield coefficient for aerobic and anaerobic processes are shown in Table 3.7. Anaerobic processes have a low yield coefficient because much of the energy in the original substrate is lost from the liquid as methane gas, with less energy available for synthesis of new cell material than in aerobic processes.

Table 3.7 Typical values of maximum yield and decay coefficients for various types of wastes treated in aerobic and anaerobic processes (Metcalf & Eddy, 1991).

Waste	Aerobic		Anaerobic	
	Y_{max}[a]	k_d[b]	Y_{max}	k_d
Domestic wastewater	0.6	0.06		
Domestic sludge			0.06	0.03
Fatty acid			0.05	0.04
Carbohydrate			0.24	0.03
Protein			0.075	0.014

[a] Y_{max} — maximum yield coefficient, lb VSS[c] produced/lb BOD_5 removed (kg/kg).

[b] k_d — decay coefficient, day^{-1}.

[c] VSS — volatile suspended solids.

Variation of the required nutrient ratio with solids retention time (defined as the total volume of solids in the digester divided by the volume of solids wasted from the system daily) is presented in Figure 3.6 for the anaerobic treatment of an acetic acid waste at 35°C (95°F). Nitrogen and phosphorus requirements also decrease with increasing solids retention time because of decreased cell production at higher than at lower sludge ages.

Requirements for calcium, potassium, magnesium, sodium, and trace elements in aerobic processes can be determined from the information provided in Table 3.8.

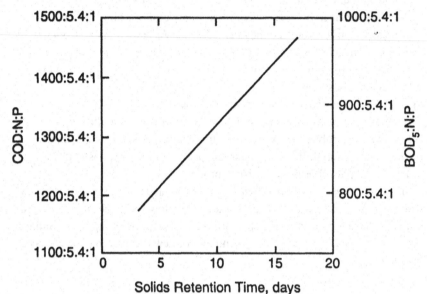

Figure 3.6 **Variation of nutrient requirements with solids retention time in anaerobic treatment processes (Broderick and Sher-**

Table 3.8 **Nutrient requirements for wastewater treatment plants (Grau, 1991).**

Nutrients	BOD_5/nutrient, lb/lb*
Calcium	100/0.62
Potassium	100/0.45
Magnesium	100/0.30
Molybdenum	100/0.043
Zinc	100/0.016
Copper	100/0.015
Cobalt	100/0.013
Sodium	100/0.005

* lb/lb × 1 000 = mg/kg.

The nutrient requirements of organisms also can be determined from their composition. The following formulas can be used for this purpose:

$$N_{req} = \left(X_{dn} \frac{X_d}{X_{di}} + X_{rn} \frac{X_n}{X_{di}} \right) P_x \qquad (3.4)$$

$$P_{req} = \left(X_{dp} \frac{X_d}{X_{di}} + X_{rp} \frac{X_n}{X_{di}} \right) P_x \qquad (3.5)$$

Where

N_{req}	=	nitrogen requirement, lb/d (kg/d);
P_{req}	=	phosphorus requirement, lb/d (kg/d);
X_{dn}	=	nitrogen content of the biodegradable fraction of cells;
X_d	=	biodegradable fraction of cells;
X_{di}	=	biodegradable fraction of cells when nutrients are not limiting;
X_{rn}	=	nitrogen content of nonbiodegradable fraction of cells;
X_n	=	nonbiodegradable fraction of cells;
X_{dp}	=	phosphorus content of biodegradable fraction of cells;
X_{rp}	=	phosphorus content of nonbiodegradable fraction of cells; and
P_x	=	amount of sludge wasted per day, lb/d (kg/d).

The amount of sludge wasted per day is calculated from the following formula:

$$P_x = Y_{obs} \, Q \, (S_0 - S_1) \, (8.34) \qquad (3.6)$$

Where

Q	=	flow rate, mgd (m^3/d);
S_0	=	influent substrate concentration, mg/L;
S_1	=	effluent substrate concentration, mg/L; and
8.34	=	conversion factor, lb/gal (lb/gal × 0.119 8 = kg/L).

The biodegradable fraction of cells at different sludge ages can be determined from the following equation:

$$X_d = \frac{X_{di}}{1 + (1 - X_{di}) \, k_d \theta_c} \qquad (3.7)$$

Where

X_d	=	biodegradable fraction of cells;

X_{di}	=	biodegradable fraction of cells when nutrients are not limiting;
k_d	=	decay coefficient, day^{-1}; and
θ_c	=	sludge age, lb solids in aeration tanks/lb solids wasted per day (kg/kg).

The values of X_{dn} and X_{dp} are given in Table 3.4 for various cell compositions. For example, X_{dn} and X_{dp} are 12.2% and 2.3%, respectively, for a cell composition of $C_{60}H_{87}O_{23}N_{12}P$. Nitrogen and phosphorus contents of non-biodegradable fraction of cells can be assumed to be 7% and 1%, respectively. The biodegradable and nonbiodegradable fraction of cells when nutrients are not limiting can be assumed as 80% and 20%, respectively (Grau, 1991). Equations 3.4, 3.5, and 3.7 reduce to Equations 3.8, 3.9, and 3.10 using the values given above.

$$N_{req} = (0.065 \, X_d + 0.07) \, P_x \tag{3.8}$$

$$P_{req} = (0.016 \, 25 \, X_d + 0.01) \, P_x \tag{3.9}$$

$$X_d = \frac{0.8}{1 + 0.2 \, k_d\theta_c} \tag{3.10}$$

Determining exact quantities of potassium, sodium, magnesium, chlorine, iron, and trace elements required for cell synthesis is more difficult. The elemental composition of organisms given in Table 3.1 can be used to estimate their requirements for growth, though such an approach may pose problems. When these nutrients are in excess of their metabolic requirements in a growth medium, they may be adsorbed by organisms. This leads to higher nutrient requirements than are actually needed for cell synthesis.

The major objective in most biological treatment processes is the reduction of carbonaceous BOD (or COD). Therefore, other nutrients are required in proportion to carbonaceous BOD. If the concentration of other nutrients is low or in excess, certain problems arise, which are addressed in the following sections.

NUTRIENT DEFICIENCY

Nutrient deficiency in wastewater or during the treatment process can have adverse effects on the process. Nutrient deficiency can lead to incomplete conversion of organic and inorganic substances to end products. Inefficient removal of BOD_5 or COD often have been reported as resulting from deficiencies in nitrogen and phosphorus. Inhibition of nitrification because of iron deficiency also has been reported.

Nutrient deficiency also can lead to inefficient removal of suspended solids. Low aeration basin dissolved oxygen (DO) concentrations were reported to cause turbid effluents in activated sludge processes (Starkey and Karr, 1984). Nutrient deficiency also can result in a shift in the microbial population of the process. Filamentous organisms can effectively use some nutrients that are in low concentrations. Consequently, they grow faster than nonfilamentous organisms under nutrient deficit conditions (Figure 3.7). A change of microbial population to filamentous bacteria, fungi, and actinomycetes (*Nocardia* and related organisms) frequently has been observed in activated sludge processes as a result of nutrient deficiency.

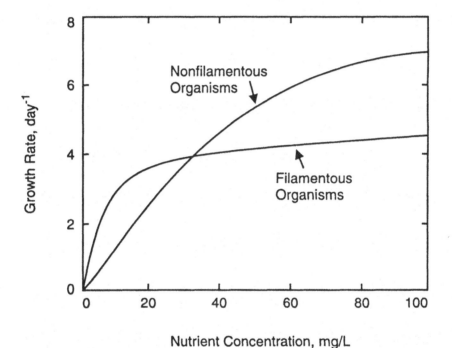

Figure 3.7   **Effect of nutrient concentration on growth rates of filamentous and nonfilamentous organisms.**

SLUDGE BULKING. When filamentous organisms are abundant, activated sludge does not settle and compact well in clarifiers, resulting in process upset. Such a condition is called *bulking*. A list of filamentous organisms responsible for bulking through nutrient deficiency is given in Table 3.9. Some of the filamentous organisms in Table 3.9 and the text are designated by number (Types 1701, 0041, and others). These organisms, which are bacteria, have not been assigned to any genera, but typically are recognized by their morphological characteristics. The reader is referred elsewhere for their identification and

---

Table 3.9   Filamentous organisms causing bulking because of nutrient deficiency in activated sludge processes.

Filamentous organisms	Deficient nutrients
Type 1701	DO*
Type 021N	N, P
Type 0041	N, P
Thiothrix sp.	N, P
Sphaerotilus natans	DO, P
Haliscomenobacter hydrossis	DO, N, P
Type 0675	N, P

* DO = dissolved oxygen.

the determination of growth conditions under which they cause bulking (Jenkins *et al.*, 1984, and WPCF, 1990).

Other nutrients such as calcium, magnesium, potassium, manganese, iron, cobalt, copper and zinc, growth factors, and vitamins also can lead to bulking when they are in deficit. Domestic wastewater is assumed to contain adequate concentrations of these nutrients. However, certain physical, chemical, and biological processes that can occur before biological wastewater treatment processes may reduce the concentration of these nutrients. When domestic wastewater is subjected to anaerobic conditions in sewers, sulfates are reduced to sulfides, which form highly insoluble metal sulfides with certain nutrients, such as ferrous iron, as shown in Equation 3.11 (U.S. EPA, 1985).

$$Fe^{2+} + HS^- \rightarrow FeS + H^+ \tag{3.11}$$

Consequently, the concentration of these nutrients can be reduced significantly, resulting in nutrient-deficient conditions.

In a similar situation, digester supernatants typically contain high concentrations of hydrogen sulfide. When digester supernatants are discharged to domestic wastewater at the head of treatment plants, these nutrients may precipitate, causing nutrient deficiency.

Wood and Tchobanoglous (1975) reported that the Sacramento County Central Treatment Plant treating primarily domestic wastewater had been troubled with the growth of filamentous organisms. The addition of industrial wastes, principally food-processing and cannery wastes, aggravated the bulking problem. The bulking was caused by a trace nutrient deficiency resulting from several factors. The carriage water contained low concentrations of trace nutrients, and the domestic wastewater had hydrogen sulfide concentrations as high as 6.5 mg/L, which might have further reduced nutrient concentrations. Food-processing and cannery wastes also had low concentrations of these nutrients, which when added to the bench- and pilot-scale processes, improved the settling rate and decreased the amount of filamentous organisms present.

Low nutrient concentrations are encountered in completely mixed activated sludge plants where incoming waste suddenly is diluted to low concentrations. Low nutrient concentrations also occur in plants operated at long sludge ages, where the concentration of biomass is high. Thus, the food available per unit weight of biomass is low. Filamentous organisms causing bulking at low food-to-microorganism (F:M) ratios include Types 021N, 0041, 0092, 0675, 0803, 0961, 0581, *Haliscomenobacter hydrossis*, *Nostocoida limicola*, and 1851.

SLUDGE FOAMING. Nutrient deficiency not only causes filamentous activated sludge bulking but also causes activated sludge foaming. Ostrander (1992) reported that *Microthrix parvicella* caused foaming and bulking problems at a treatment plant during nitrification because of the deficiency of a nutrient. Problems were solved by supplying a vitamin (folic acid) to the secondary influent, significantly reducing the *M. parvicella* population. Other foam-causing filamentous organisms frequently observed in activated sludge plants include *Nocardia* spp., *Rhodococcus* spp., and Type 1863 (Sezgin *et al.*, 1988, and Nowak, 1986). Both *Nocardia* and *Rhodococcus* occur at low F:M conditions, and Type 1863 occurs at low DO combined with high F:M.

Some industrial wastes are deficient in nitrogen or phosphorus, as listed in Table 3.10. The BOD:N:P ratio for the treatment of nutrient-deficient wastewater and the form of chemical addition applied to these wastes are listed in Table 3.11 for different types of processes.

*N*UTRIENT ADDITION

The existence of nutrient-deficient conditions in a process can be detected from plant operating conditions. Lower BOD_5 removals (or high-effluent BOD_5 levels) and deterioration of settling characteristics of activated sludge may indicate nutrient deficiency. To confirm nutrient deficiency and determine which nutrients are deficient, an analysis of secondary influent can be conducted for BOD_5 (or COD), nitrogen forms (ammonia and organic nitrogen), phosphorus, and other nutrients listed in Tables 3.1 and 3.2.

The required ratio of BOD_5:N:P for optimum treatment efficiency can be determined from Figure 3.3 for a given sludge age, specific yield, and decay coefficients for an activated sludge process. If the BOD_5:N:P ratio found from the analysis of the secondary influent is higher than that found from Figure 3.3, a nutrient deficiency may exist. Comparison of BOD_5 to nutrient ratios found from the analysis of secondary influent with those ratios provided in Table 3.8 may indicate whether other nutrients are deficient. If nutrient deficiency is confirmed, the amount of a particular nutrient needed should be determined using one of the methodologies described above. The amount of

Table 3.10 Nutrient-deficient industrial wastes (Broderick and Sherrard, 1985).

Waste type	Deficient nutrient
Bakery	N
Bottling plant	N, P
Brewery	N
Citrus	N
Chemical plant	P
Coffee, soluble	N
Coke oven	P
Corn	N
Cotton keiring	N
Dairy	N, P
Food processing	N, P
Formaldehyde	N, P
Fruits and vegetables	N, P
Paper and pulp	N, P
Pear	N, P
Pharmaceutical	P
Phenols	N
Pineapple	N, P
Rag and rope	N, P
Sugar beets	N
Soybean	N
Textile	N
Vinegar	N, P
Winery	N, P

nutrient available is found from the analysis of secondary influent. If the difference between nutrient needed and nutrient available is zero or a negative number, there is no shortage. If the difference is a positive number, the particular nutrient should be added in that amount.

After the addition of the required nutrient has started, effluent quality and settling characteristics of activated sludge should be monitored. If an improvement of effluent quality (BOD_5 or COD removal) and settling characteristics is not observed after a period of operation corresponding to three sludge ages, the dose of the deficient nutrient should gradually be increased until improvements are observed in effluent quality and sludge-settling characteristics. If the nutrient dose selected originally improves both effluent quality and sludge-settling characteristics, the dose can be lowered to help minimize chemical costs.

Table 3.11   BOD$_5$:N:P ratios for various industrial wastes fed to activated sludge processes (Broderick and Sherrard, 1985).

Waste type	BOD$_5$:N:P	Form of addition
Bleach	100:4.45:0.95	Anhydrous NH_3, $(NH_4)_2HPO_4$
Bleach	100:2.8:0.8	
Brewery	20:1:-	
Brewery	100:4.2:1.04	
Citrus	100:5:1	Anhydrous NH_3
Coffee	60:3:1	Bactopeptone, NH_4NO_3
Coke oven	100:4.8	Na_2HPO_4
Cotton kiering	100:4.2:0.54	
Fiberboard	100:5:1	Anhydrous NH_3, H_3PO_4
Formaldehyde	100:5:0.67	Anhydrous NH_3, $(NH_4)_2HPO_4$
Fruits and vegetables	100:2:0.5	
Paper	100:4:1	
Rag and rope	100:2.2:0.44	
Soybean	100:4:1	
Vinegar	100:8:2.66	NH_4Cl, $(NH_4)_2HPO_4$

NUTRIENT DOSE ON THE BASIS OF THE RATIO OF INFLUENT BIOCHEMICAL OXYGEN DEMAND TO NUTRIENT WEIGHT. The amount of a specific nutrient needed is determined from the following formula:

$$\text{Nutrient needed, } {}^{mg}\!/_L = \frac{\text{Secondary influent BOD}_5, {}^{mg}\!/_L}{\text{Weight ratio, } {}^{BOD_5}\!/_{nutrient}} \qquad (3.12)$$

The nutrient shortage, if it exists, is found as the difference between the amount of nutrient needed and nutrient available.

$$\text{Nutrient shortage, } {}^{mg}\!/_L = \text{Nutrient needed, } {}^{mg}\!/_L - \text{Nutrient available, } {}^{mg}\!/_L \qquad (3.13)$$

and

$$\text{Nutrient to add, } {}^{lb}\!/_d = \text{Nutrient shortage, } {}^{mg}\!/_L \text{ (flow rate, mgd) } (8.34 \, {}^{lb}\!/_{gal}) \qquad (3.14)$$

A list of the nitrogen and phosphorus compounds that can be added as nutrient supplements to biological treatment processes is presented in Table 3.12. Compounds used as supplements for other nutrients are given in Table 3.13.

Table 3.12 Nitrogen and phosphorus compounds suitable for addition to biological treatment processes (Gerardi, 1991).

Source	Name/formula
Nitrogen	Anhydrous ammonia, NH_3; ammonium hydroxide, NH_4OH; ammonium bicarbonate NH_4HCO_3; ammonium carbonate, $(NH_4)_2CO_3$; ammonium chloride, NH_4Cl, ammonium phosphate, $NH_4H_2PO_4$; ammonium sulfate, $(NH_4)_2SO_4$
Phosphorus	Trisodium phosphate, Na_3PO_4; disodium phosphate, Na_2HPO_4; monosodium phosphate, NaH_2PO_4; ammonium phosphate, $NH_4H_2PO_4$; sodium hexametaphosphate, $Na_3(PO_3)_6$; sodium tripolyphosphate, $Na_5P_3O_{10}$; tetrasodium pyrophosphate, $Na_4P_2O_7$; phosphoric acid, H_3PO_4

Table 3.13 Compounds suitable for addition to biological treatment processes (Neidhardt *et al.*, 1990).

Nutrient	Name/compound
Sulfur	Sodium sulfate, Na_2SO_4; hydrogen sulfide, H_2S; ammonium sulfate, $(NH_4)_2SO_4$
Calcium	Calcium chloride, $CaCl_2$
Potassium	Potassium chloride, KCl; potassium phosphate, dibasic, K_2HPO_4
Magnesium	Magnesium chloride, $MgCl_2$; magnesium sulfate, $MgSO_4$
Sodium	Sodium chloride, $NaCl$
Iron	Ferric chloride, $FeCl_3$; ferric ammonium sulfate, $FeNH_4(SO_4)_2$; ferrous sulfate, $FeSO_4$; iron chelates
Trace elements	Cobalt chloride, $CoCl_2$; zinc chloride, $ZnCl_2$; sodium molybdate, Na_2MoO_4; cupric chloride, $CuCl_2$; cupric sulfate, $CuSO_4$; manganese sulfate, $MnSO_4$; nickel chloride, $NiCl_2$; sodium selenate, Na_2SeO_4; sodium tungstate, Na_2WO_4; sodium vanadate, Na_2VO_4

The chemical dose to be added can be determined from the following formula:

$$\text{Chemical, } lb/d = \frac{(\text{Nutrient to add, } lb/d) \text{ (Weight ratio) (100\%)}}{\text{Concentration of chemical, \%}} \qquad (3.15)$$

Weight ratio is determined from the ratio of molecular weight of the chemical to atomic weight of nutrient. For example, the weight ratio of $NH_3/N = 17/14 = 1.2$.

The chemical dose computed in terms of lb/d can be converted to chemical feed rate expressed as gpd (m^3/d) by Equation 3.16:

$$\text{Chemical feed rate, gpd} = \frac{\text{Chemical to add, } lb/d}{\text{Chemical density, } lb/gal} \qquad (3.16)$$

Nutrients should be fed to secondary influent in proportion to wastewater flow rate or to mass wastewater strength flow rate (lb BOD_5/d).

Example 1—calculate the amount of nutrients to be added to wastewater treated in an activated sludge system.

Given:
— Average daily plant flow, $Q = 10$ mgd;
— Maximum yield coefficient, $Y_{max} = 0.6$ lb cells produced/lb BOD_5 removed
— Decay coefficient, $k_d = 0.05$ day^{-1};
— Sludge age, $\theta_c = 10$ days;
— Secondary influent $BOD_5 = 250$ mg/L;
— Secondary effluent BOD_5 is negligible;
— Secondary influent ammonia-nitrogen = 4 mg/L;
— Secondary influent organic nitrogen = 2 mg/L (70% of organic nitrogen is available);
— Secondary influent phosphorus = 1.5 mg/L;
— Secondary influent iron = 0.75 mg/L;
— Secondary influent zinc = 0.01 mg/L;
— Chemicals available: ammonia (NH_3); trisodium phosphate (Na_3PO_4); ferric chloride ($FeCl_3$); and zinc chloride ($ZnCl_2$);
— Ammonia-to-nitrogen atomic weight ratio, $NH_3:N = 17:14 = 1.2$;
— Trisodium phosphate:phosphorus atomic weight ratio, $Na_3PO_4:P = 164:31 = 5.3$;
— Ferric chloride:iron atomic weight ratio, $FeCl_3:Fe = 162.5:56 = 2.9$;
— Zinc chloride:zinc atomic weight ratio, $ZnCl_2:Zn = 2.1$; and
— Density of 80% ammonia solution: 1.80 kg/L (15.02 lb/gal); density of 75% trisodium phosphate solution: 1.75 kg (14.6 lb/gal);

density of 30% ferric chloride solution: 1.30 kg/L (10.84 lb/gal); and density of 20% zinc chloride solution: 1.20 kg/L (10 lb/gal).

Solution:

Step 1

— Use Figure 3.3 to determine BOD_5:N:P ratio for conditions of $Y_{max} = 0.6$, $k_d = 0.05$, and $\theta_c = 10$ days. A BOD_5:N:P ratio of 110:5.4:1 is found from Figure 3.3.

— BOD_5:N = 110:5.4 = 20.4 and BOD_5:P = 110:1 = 110.

— Choose a BOD_5:Fe ratio of 100:0.5 = 200 (WPCF, 1987) and BOD_5:Zn = 100:0.016 = 6 250 from Table 3.8.

Step 2

— Calculate the amount of nutrient needed to achieve the required BOD_5:nutrient ratios using Equation 3.12.

$$\text{Nitrogen needed, } \text{mg}/\text{L} = \frac{\text{Secondary influent BOD}_5, \text{mg}/\text{L}}{\text{Required ratio, BOD}_5\text{:N}} = \frac{250}{20.4} = 12.25$$

$$\text{Phosphorus needed, } \text{mg}/\text{L} = \frac{\text{Secondary influent BOD}_5, \text{mg}/\text{L}}{\text{Required ratio, BOD}_5\text{:P}} = \frac{250}{110} = 2.27$$

$$\text{Iron needed, } \text{mg}/\text{L} = \frac{\text{Secondary influent BOD}_5, \text{mg}/\text{L}}{\text{Required ratio, BOD}_5\text{:Fe}} = \frac{250}{200} = 1.25$$

$$\text{Zinc needed, } \text{mg}/\text{L} = \frac{\text{Secondary influent BOD}_5, \text{mg}/\text{L}}{\text{Required ratio, BOD}_5\text{:Zn}} = \frac{250}{6\ 250} = 0.04$$

Step 3

— Calculate nutrient shortage (from Equation 3.13) as the difference between nutrient needed minus nutrient available. If the result is zero or a negative number, no nutrient shortage exists.

Nitrogen available for nutritional use $= (\text{Ammonia–nitrogen, mg}/\text{L}) + (0.70) (\text{Organic nitrogen, mg}/\text{L})$

$$= 4 + (0.70) (2)$$
$$= 5.4 \text{ mg}/\text{L}$$

Nitrogen shortage, $\text{mg}/\text{L} = (\text{N needed, mg}/\text{L}) - (\text{N available, mg}/\text{L})$
$$= 12.25 - 5.4$$
$$= 6.85 \text{ mg}/\text{L}$$

Phosphorus shortage, mg/L = (P needed, mg/L) − (P available, mg/L)

$$= 2.27 - 1.5$$
$$= 0.77 \text{ mg/L}$$

Iron shortage, mg/L = (Fe needed, mg/L) − (Fe available, mg/L)

$$= 1.25 - 0.75$$
$$= 0.5 \text{ mg/L}$$

Zinc shortage, mg/L = (Zn needed, mg/L) − (Zn available, mg/L)

$$= 0.04 - 0.01$$
$$= 0.03 \text{ mg/L}$$

Step 4

— Calculate the weight of nutrients that need to be added (from Equation 3.14).

Nitrogen to add, lb/d = (N shortage, mg/L) (Q, mgd)(8.34, lb/gal)

$$= (6.85)\,(10)\,(8.34)$$
$$= 571 \text{ lb/d } (259 \text{ kg/d})$$

Phosphorus to add, lb/d = (P shortage, mg/L) (Q, mgd)(8.34, lb/gal)

$$= (0.77)\,(10)\,(8.34)$$
$$= 64.2 \text{ lb/d } (29.1 \text{ kg/d})$$

Iron to add, lb/d = (Fe shortage, mg/L) (Q, mgd)(8.34, lb/gal)

$$= (0.5)\,(10)\,(8.34)$$
$$= 41.7 \text{ lb/d } (18.9 \text{ kg/d})$$

Zinc to add, lb/d = (Zn shortage, mg/L) (Q, mgd)(8.34, lb/gal)

$$= (0.03)\,(10)\,(8.34)$$
$$= 2.50 \text{ lb/d } (1.1 \text{ kg/d})$$

Step 5

— Calculate the weight of the commercial chemical to be added per day.
— A commercial grade solution with an 80% concentration of anhydrous ammonia will be used as the nitrogen source.

$$\text{Anhydrous ammonia, lb/d} = \frac{(\text{Nitrogen to add, lb/d})\,(\text{Weight ratio of } NH_3\!:\!N)\,(100\%)}{NH_3 \text{ concentration, }\%}$$

$$= \frac{(571)\,(1.2)\,(100)}{80}$$
$$= 856 \text{ lb/d } (388 \text{ kg/d})$$

— A commercial grade solution with a 75% concentration of trisodium phosphate will be used as the phosphorus source.

$$\text{Trisodium phosphate, } ^{lb}\!/_d = \frac{(\text{Phosphorus to add, } ^{lb}\!/_d)(\text{Weight ratio of } Na_3PO_4\!:\!P)(100\%)}{Na_3PO_4 \text{ concentration, \%}}$$

$$= \frac{(64.2)(5.3)(100)}{75}$$

$$= 454 \; ^{lb}\!/_d \; (206 \; ^{kg}\!/_d)$$

— A commercial grade solution with a 30% concentration of ferric chloride will be used as the iron source.

$$\text{Ferric chloride, } ^{lb}\!/_d = \frac{(\text{Iron to add, } ^{lb}\!/_d)(\text{Weight ratio of } FeCl_3\!:\!Fe)(100\%)}{FeCl_3 \text{ concentration, \%}}$$

$$= \frac{(41.7)(2.9)(100)}{30}$$

$$= 403 \; ^{lb}\!/_d \; (183 \; ^{kg}\!/_d)$$

— Zinc chloride will be prepared as a 20% solution.

$$\text{Zinc chloride, } ^{lb}\!/_d = \frac{(\text{Zinc to add, } ^{lb}\!/_d)(\text{Weight ratio of } ZnCl_2\!:\!Zn)(100\%)}{ZnCl_2 \text{ concentration, \%}}$$

$$= \frac{(2.5)(2.1)(100)}{20}$$

$$= 26 \; ^{lb}\!/_d \; (12 \; ^{kg}\!/_d)$$

Step 6

— Calculate the chemical feed rate. Use Equation 3.16 to calculate the chemical feed rate.

$$\text{Anhydrous ammonia feed rate, gpd} = \frac{\text{Anhydrous ammonia to add, } ^{lb}\!/_d}{\text{Anhydrous ammonia density, } ^{lb}\!/_{gal}}$$

$$= \frac{856}{15.02}$$

$$= 57 \text{ gpd } (0.22 \; m^3\!/_d)$$

$$\text{Trisodium phosphate feed rate, gpd} = \frac{\text{Trisodium phosphate to add, } ^{lb}\!/_d}{\text{Trisodium phosphate density, } ^{lb}\!/_{gal}}$$

$$= \frac{45}{14.60}$$

$$= 31.1 \text{ gpd } (0.12 \; m^3\!/_d)$$

$$\frac{\text{Ferric chloride}}{\text{feed rate, gpd}} = \frac{\text{Ferric chloride to add, } ^{lb}\!/_d}{\text{Ferric chloride density, } ^{lb}\!/_{gal}}$$

$$= \frac{403}{10.84}$$

$$= 37.2 \text{ gpd } (0.14 \text{ m}^3\!/_d)$$

$$\frac{\text{Zinc chloride}}{\text{feed rate, gpd}} = \frac{\text{Zinc chloride to add, } ^{lb}\!/_d}{\text{Zinc chloride density, } ^{lb}\!/_{gal}}$$

$$= \frac{26}{10}$$

$$= 2.6 \text{ gpd } (0.01 \text{ m}^3\!/_d)$$

NUTRIENT DOSE ON THE BASIS OF CELL COMPOSITION. The amount of a nutrient needed by organisms is calculated from Equations 3.3 through 3.7.

Example 2—Calculate the nutrient dose to be added to a wastewater treated in an activated sludge process.

Given:
- Average daily plant flow, Q = 10 mgd;
- Maximum yield coefficient, Y_{max} = 0.6 lb cells produced/lb BOD_5 removed;
- Decay coefficient, k_d=0.05 day^{-1};
- Sludge age, θ_c = 10 days;
- Secondary influent BOD_5 = 250 mg/L;
- Secondary effluent soluble BOD_5 = 7 mg/L;
- Secondary influent ammonia-nitrogen: 4 mg/L;
- Secondary influent organic nitrogen = 2 mg/L (70% of organic nitrogen is available to organisms for nutritional use);
- Secondary influent phosphorus = 1.5 mg/L; and
- Cell composition is $C_{60}H_{87}O_{23}N_{12}P$. Nonbiodegradable fraction of cells has 7% nitrogen and 1% phosphorus.

Solution:

Step 1
- Calculate the biodegradable fraction of cells using Equation 3.10.

$$X_d = \frac{0.8}{1 + (0.2)\,(0.05)\,(10)} = 0.73$$

Step 2

— Calculate the observed yield coefficient at sludge age of 10 days from Equation 3.

$$Y_{obs} = \frac{0.6}{1 + (0.05)(10)} = 0.40 \text{ lb cell produced}/\text{lb BOD}_5 \text{ removed}$$

Step 3

— Calculate the amount of sludge (cells or VSS) produced per day from Equation 3.6.

$$P_x \text{ lb/d} = (0.4 \text{ lb cells}/\text{lb BOD}_5)(10 \text{ mgd})(250 - 7, \text{mg/L})(8.34, \text{lb/gal})$$
$$= 8\ 106 \text{ lb/d} (3\ 676 \text{ kg/d})$$

Step 4

— Calculate the nitrogen and phosphorus requirements from Equations 3.8 and 3.9.

Nitrogen needed, $\text{lb/d} = [(0.065)(0.73) + 0.07](8\ 106 \text{ lb/d})$
$$= 952 \text{ lb/d} (432 \text{ kg/d})$$

Phosphorus needed, $\text{lb/d} = [(0.016\ 25)(0.73) + 0.01](8\ 106 \text{ lb/d})$
$$= 177 \text{ lb/d} (80 \text{ kg/d})$$

Step 5

— Calculate nutrients available per day, lb/d.

Nitrogen available for nutritional use $= (\text{Ammonia–nitrogen, mg/L}) + (0.70)(\text{Organic nitrogen, mg/L})$
$$= 4 + (0.70)(2)$$
$$= 5.4 \text{ mg/L}$$

Nitrogen available, $\text{lb/d} = (5.4 \text{ mg/L})(10 \text{ mgd})(8.34 \text{ lb/gal})$
$$= 450 \text{ lb/d} (204 \text{ kg/d})$$

Phosphorus available, $\text{lb/d} = (1.5 \text{ mg/L})(10 \text{ mgd})(8.34 \text{ lb/d})$
$$= 125 \text{ lb/d} (57 \text{ kg/d})$$

Step 6

— Calculate the amount of nutrient to add.

Nitrogen to add, $\text{lb/d} = (\text{N needed, lb/d}) - (\text{N available, lb/d})$
$$= 952 - 450$$
$$= 502 \text{ lb/d} (228 \text{ kg/d})$$

$$\text{Phosphorus to add, } ^{lb}\!/_d = (\text{P needed, } ^{lb}\!/_d) - (\text{P available,} ^{lb}\!/_d)$$
$$= 177 - 125$$
$$= 52 \ ^{lb}\!/_d \ (24 \ ^{kg}\!/_d)$$

Calculation of nutrient doses and chemical feed rates in Example 2 is the same as that given in Example 1.

Nitrogen can be added in the form of organic (urea) or inorganic nitrogen. Because organic nitrogen exerts an additional BOD on the process and is only partially available to organisms as a nitrogen source, nitrogen typically is added in an inorganic form, either NH_4^+ or NO_3^-. The advantages of adding nitrogen, either NH_4^+ or NO_3^-, are listed in Table 3.14. Ammonia can be added to plants in which nitrification is not required and is used as a control parameter for nutrient addition. Ammonia effluent concentration should be 1.5 mg/L or higher.

Table 3.14   Comparison of the effects of ammonia and nitrate addition on activated sludge processes (Emerson and Sherrard, 1983).

NH_4^+–N addition	NO_3–N addition
Higher sludge production	Lower sludge production
Higher oxygen requirements	Lower oxygen requirements
Destruction of alkalinity	Production of alkalinity
Higher MLVSS[a]	Lower MLVSS
Better COD[b] removal	Poorer COD removal

[a]   MLVSS = mixed liquor volatile suspended solids.

[b]   COD = chemical oxygen demand.

In plants where nitrification or nitrogen removal is required, both ammonia and nitrate can provide nitrogen sources. Because ammonia is directly incorporated to the cell material without reduction, it is depleted first during nutritional use. The remaining nitrogen requirement is satisfied with the nitrate dosing. Ammonia, for example, can be supplied at 80% of the nitrogen requirement, with the remaining 20% of the nitrogen requirement being satisfied by nitrate dosing. Nitrates remaining in the effluent can be removed by endogenous denitrification in a relatively small anoxic zone at the outlet end of the aeration tank (Grau, 1991).

Phosphorus is added in the form of orthophosphate, which is readily usable by organisms. Soluble phosphorus or orthophosphate concentration in the effluent can be used as a control criteria for nutrient addition. A soluble phosphorus concentration of 0.5 mg/L in the effluent indicates that the phosphorus requirement of organisms is satisfied in the aeration tank. In many cases, however, even lower concentrations of soluble phosphorus have been reported without adverse effects on sludge settleability or effluent quality.

Nutrients also can be supplied through the addition of digester supernatant in some treatment plants, such as those employing the Kraus process, which is used to treat wastewater with low nitrogen levels. A portion of return sludge is mixed with digester supernatant and solids in a nitrifying aeration tank, and the resulting mixed liquor is added to the return sludge and mixed in the aeration tank.

OXYGEN REQUIREMENTS OF ORGANISMS

OXYGEN REQUIREMENT FROM THE RATIO OF OXYGEN TO BIOCHEMICAL OXYGEN DEMAND. The oxygen requirement of organisms can be determined from their growth equation, written for the oxidation of substrate and the synthesis of new cell material. Figure 3.5 has been developed using the growth equation for the case of Y_{max} = 0.5 lb VSS/lb BOD_5 and k_d = 0.06 day^{-1}. Similar curves can be developed for other kinetic constants. The use of Figure 3.5 is illustrated with Example 3.

Example 3—calculate oxygen requirement in an activated sludge process.

Given:
- — Influent flow rate, Q = 10 mgd;
- — Influent BOD_5, S_0 = 150 mg/L;
- — Effluent soluble BOD_5, S_1 is negligible;
- — Sludge age, θ_c = 10 days;
- — Maximum yield coefficient, Y_{max} = 0.5 lb VSS produced/lb BOD_5 removed;
- — Decay coefficient, k_d = 0.06 day^{-1}; and
- — The ratio of ultimate biochemical oxygen demand (BOD_u) to BOD_5 = 1.5.

Solution:

Step 1
- — Determine the ratio of lb O_2 required/lb BOD_5 removed from Figure 3.5. For θ_c = 10 days, lb O_2/lb BOD_5 = 1.08.

Step 2
- — Calculate the oxygen requirement from the ratio of lb O_2/lb BOD_5.

Oxygen requirement, lb/d = (1.08) (BOD$_5$ removal)

BOD$_5$ removal, lb/d = (flow rate, mgd) (influent BOD$_5$, mg/L) (8.34, lb/gal)

$$= (10)(150)(8.34)$$
$$= 12\ 510\ lb/d\ (5\ 675\ kg/d)$$

Oxygen requirement, lb/d = (1.08) (12 510)
$$= 13\ 511\ lb/d\ (6\ 129\ kg/d)$$

OXYGEN REQUIREMENT FOR AN ACTIVATED SLUDGE PROCESS WITHOUT NITRIFICATION. The oxygen requirement for an activated sludge process without nitrification can be determined from the amount of carbonaceous substrate oxidized per day. However, not all substrate is oxidized, but a portion is used for synthesis of new cell material. At steady state, the amount of new cells synthesized per day is equal to the amount of cells (or sludge) wasted per day. Therefore, the oxygen equivalent of biomass produced per day should be subtracted from the amount of oxygen required to oxidize the substrate to determine the total oxygen requirement. The oxygen equivalent of biomass can be found from the following equation assuming a cell composition of $C_5H_7O_2N$:

$$C_5H_7O_2N\ +\ \ \ \ 5O_2\ \ \ \ \to 5CO_2 + 2H_2O + NH_3$$

113(mol wt)   5×(32 mol wt)   (3.17)

Thus, 1 mg of the organic fraction of biomass exerts an oxygen demand of 1.42 mg ($5 \times 32/113 = 1.42$).

$$\text{Oxygen requirement, } lb/d = \frac{\text{substrate}}{\text{oxidized, } lb/d} - 1.42 \frac{\text{organisms}}{\text{wasted, } lb/d} \quad (3.18)$$

$$\frac{\text{Substrate}}{\text{oxidized, } lb/d} = \text{(flow rate, mgd) (substrate removed, } mg/L)\ (8.34,\ lb/gal) \quad (3.19)$$

The amount of organisms wasted per day can be found from Equation 3.6.

Example 4—calculate the oxygen requirement for an activated sludge process.

Given: The same operating conditions as stated in Example 3.

Solution:

Step 1
— Calculate the amount of substrate oxidized per day from Equation 3.19.

$$\text{Substrate oxidized, } ^{lb}\!/_d = (Q, \text{mgd}) (S_0 - S_1, \, ^{mg}\!/_L) (^{BOD_u}\!/_{BOD_5}) (8.34, \, ^{lb}\!/_{gal})$$

$$= (10) (150 - 0) (1.5) (8.34)$$

$$= 18\ 765 \, ^{lb}\!/_d \, (8\ 511 \, ^{kg}\!/_d)$$

Step 2
— Calculate the amount of organisms wasted per day. The observed yield coefficient is determined from Equation 3.3.

$$\text{Observed yield, } ^{lb\ VSS}\!/_{lb\ BOD_5} \text{ coefficient} = \frac{0.5}{1 + (0.06)\ (10)}$$

$$= 0.31 \, ^{lb\ VSS}\!/_{lb\ BOD_5}$$

The amount of organisms wasted per day is calculated from Equation 3.6.

$$\text{Organisms wasted, } ^{lb\ VSS}\!/_d = (0.31)\ (10)\ (150 - 0)\ (8.34)$$

$$= 3\ 909 \, ^{lb\ VSS}\!/_d \, (1\ 773 \, ^{kg}\!/_d)$$

Step 3
— Calculate the amount of oxygen required per day from Equation 3.18.

$$\text{Oxygen required, } ^{lb}\!/_d = 18\ 765 - (1.42)\ (3\ 909)/$$

$$= 13\ 214 \, ^{lb}\!/_d \, (5\ 994 \, ^{kg}\!/_d)$$

Excess Nutrients

Excessive discharge of nutrients can have adverse effects on the environment. Large quantities of carbon (BOD) release, for example, from bypass of raw wastewater under certain circumstances can reduce DO content of the receiving water, leading to fish kills. High concentrations of nitrogen and phosphorus result in overfertilization of rivers, lakes, and estuaries, leading to excessive plant growth or algae blooms. This causes DO depletion and fish kills in receiving waters. An algae bloom can lead to taste and odor problems and can increase turbidity and add color to the water, which becomes unsightly and unsuitable for recreational and other uses.

Ammonium discharged to receiving water exerts additional oxygen demand because of its oxidation. An effluent with 95% BOD_5 removal in a conventional wastewater treatment plant can exert more than 100 mg/L of oxygen demand, assuming an influent BOD_5 of 250 mg/L and influent Kjeldahl nitrogen of 40 mg/L; but, if ammonia oxidation takes place in the plant, the total oxygen demand of the effluent can be reduced to less than 40 mg/L.

Ammonium ions in receiving waters can be converted to ammonia with an increase in pH above 7.0 (Equation 3.2). Ammonia is toxic to fish and many other aquatic organisms in concentrations of less than 1.0 mg/L. Ammonia toxicity increases with an increase in DO and carbon dioxide concentrations, temperature, and bicarbonate alkalinity (U.S. EPA, 1975).

Ammonium ions reduce the disinfection efficiency of chlorine. They react with chlorine to form chloramines, which are less-effective disinfectants than free available chlorine (the sum of molecular chlorine, Cl_2; hypochlorous acid, $HOCl$; and hypochlorite ion, OCl^-). Free available chlorine exists only after breakpoint chlorination point is reached. At breakpoint chlorination, ammonium ions are completely oxidized to nitrogen gas.

As large quantities of chlorine (10 mg/L chlorine for every 1 mg/L ammonia) are required to reach the breakpoint, breakpoint chlorination seldom is practiced in wastewater treatment. Instead, chloramines are used for disinfection. In plants where partial nitrification of ammonia is achieved, nitrites in effluents can be a problem. Nitrite is oxidized to nitrate by chlorine at breakpoint chlorination. Chlorine then is reduced to chloride ions, losing its disinfecting property.

High concentrations of nitrates in water supplies can cause *methemoglobinemia*, a blood disorder in infants fewer than 3 months old. Nitrate is converted to nitrite in the stomach after digestion, which then combines with hemoglobin in the blood to form methemoglobin. This reduces the oxygen-carrying capacity of the blood, leading to suffocation and a bluish skin color.

Nitrite in the intestine can also react with amines to form *nitrosamines*, which are carcinogenic substances. U.S. EPA has placed limits on the allowable concentrations of inorganic nitrogen compounds in various types of receiving waters. Maximum allowable concentrations of NH_3, NO_3^-, and NO_2^- in public water supplies are 0.5, 10, and 1 mg/L, respectively.

High sulfide levels in wastewater can have harmful effects. Hydrogen sulfide causes odor in treatment plants, with a threshold concentration in water for detection by human smell ranging between 0.000 01 and 0.000 1 mg/L (U.S. EPA, 1974). Hydrogen sulfide also is highly toxic and can cause headache, nausea, eye injury, loss of sense of smell, edema, aprea, and even death at concentrations of 300 mg/L by volume in air. Such concentrations can develop from wastewater containing 2 mg/L of dissolved sulfide at a pH of 7.0

in an enclosed atmosphere. Hydrogen sulfide also is a precursor to the formation of sulfuric acid:

$$H_2S + 2O_2 \rightarrow H_2SO_4 \qquad (3.20)$$

Sulfuric acid corrodes concrete; metals such as iron, zinc, copper, lead, and cadmium; lead-based paints; and other materials (U.S. EPA, 1985). Concrete or metal sewer lines carrying wastewater containing high hydrogen sulfide concentrations are vulnerable to sulfuric acid corrosion.

High sulfide concentrations in wastewater lead to sludge-bulking problems in wastewater treatment plants. *Beggiatoa* and *Thiothrix*, which are filamentous bulking organisms, grow well in the presence of high sulfide concentrations. During the oxidation of hydrogen sulfide, they deposit elemental sulfur, which is used as the source of energy when hydrogen sulfide is depleted in the medium.

High concentrations of sodium, potassium, calcium, magnesium, iron, manganese, and chlorides in wastewater are objectionable because they interfere with biological treatment processes. When concentrations of these nutrients are higher outside a cell than inside, the water in the cell flows out, dehydrating the cell. This causes the *protoplast* (the membrane surrounding the cell contents) to collapse, leading to cell death. This process is known as *plasmolysis*. In addition, when these elements are discharged to receiving waters, they make the water unfit for various beneficial uses.

High concentrations of sodium, potassium, and chlorides impart a salty taste to water, making it unsuitable for drinking. Iron and manganese cause turbidity in receiving waters, impart objectionable stains to plumbing fixtures, and cause difficulties in water supply distribution systems by supporting growths of iron bacteria. Iron also gives a taste to water. Calcium, magnesium, ferrous iron, and manganese cause hardness in water, thus increasing soap consumption and producing scale in hot water pipes, heaters, and boilers.

High concentrations of copper, nickel, selenium, and zinc are toxic to both aquatic organisms and humans. Receiving waters high in these nutrients are unfit as potable water supplies.

CARBON REMOVAL

Biological processes are used to remove biodegradable organic carbon in colloidal or dissolved form. In the process, organic carbon is converted to gases such as carbon dioxide that escape to the atmosphere and to biomass that can be removed by settling. Biological treatment processes can remove 80 to 95% of influent BOD_5.

Natural treatment (self-purification) systems are used to remove carbon and other nutrients by means of physical, chemical, and biological processes occurring between water, soil, plants, organisms, and the atmosphere. Natural treatment systems include soil-based or land-treatment systems and aquatic-based systems. The typical processes that can occur in natural treatment systems include sedimentation, filtration, gas transfer, adsorption, ion exchange, chemical precipitation, chemical oxidation and reduction, biological conversion and degradation, photosynthesis, photooxidation, and plant uptake. Removal of trace organic chemicals typically is greater than 99% in natural treatment systems (Metcalf & Eddy, 1991).

NITROGEN REMOVAL

The biological removal processes used to remove nitrogen are ammonification followed by nitrification and denitrification. In ammonification, which is carried out by heterotrophic organisms, organic nitrogen (proteins and peptides) is decomposed to ammonia or ammonium ions. In nitrification, ammonia is oxidized to nitrite and then to nitrate. Oxidation is carried out by a group of autotrophic bacteria. Nitrate formed during nitrification is removed from wastewater by heterotrophic organisms through conversion to gaseous nitrogen species through *denitrification*. In this process, nitrate first is reduced to nitrite and then to nitric oxide (NO), followed by nitrous oxide (N_2O) and nitrogen gas (N_2).

Other biological nitrogen removal processes include algae harvesting, bacterial assimilation, and natural treatment. Nitrogen is incorporated to cells, which then are removed from the waste stream through algae harvesting and bacterial assimilation. These processes, however, require a carbon source, such as ethanol or glucose, to be added to the wastewater. Nitrogen removal efficiencies of 50 to 80% can be achieved with algae harvesting and 30 to 70% with bacterial assimilation. Among the natural treatment systems, irrigation can achieve 40 to 90% removal of influent nitrogen, and infiltration or percolation can achieve 0 to 50% removal (U.S. EPA, 1975).

PHOSPHORUS REMOVAL

In conventional biological treatment processes, organisms use phosphorus during cell synthesis, maintenance, and energy transport. As a result, 10 to 30% of the influent phosphorus is consumed by organisms for their metabolic processes. However, certain organisms are capable of removing phosphorus from wastewater in excess of their requirements for growth. The process of removing phosphorus beyond the metabolic requirements of organisms is known as "luxury uptake." The organisms which carry out luxury uptake of phosphorus include *Acinetobacter*, *Pseudomonas*, and *Moraxella* (Toerien *et al.*, 1990). They are collectively called poly-P bacteria because they store large amounts of phosphorus in the form of polyphosphate (metachromatic) granules, which are used as energy sources under stressed conditions. When these organisms are stained, polyphosphate granules appear as black or bluish spots on organisms under the microscope, depending on the type of stain applied.

When activated sludge is subjected to anaerobic growth conditions, some organic matter in wastewater is converted to fatty acids by acidogenic bacteria through fermentation and oxidation. The fatty acids are then taken up by the poly-P bacteria and stored in the form of polyhydroxybutyrate (PHB). The energy required for this uptake and storage is provided by the hydrolysis of poly-P reserves, which results in phosphorus release in the anaerobic stage. During the aerobic phase the stored PHB granules are used as sources of energy and for cell synthesis. The energy produced is then used to reconstitute poly-P reserves (Figure 3.8).

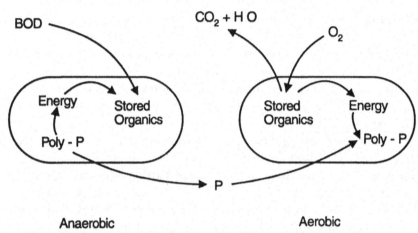

Figure 3.8   Luxury uptake of phosphorus.

Factors appearing to affect phosphorus release and uptake include substrate composition and concentration, nitrate and sulfide concentrations, and DO level. The rate of phosphorus release and uptake is a multivariant function of all of these factors. Shorter chain fatty acids such as formic, acetic, and propionic or their salts are capable of inducing phosphorus release under anaerobic, anoxic, and aerobic conditions (Gerber *et al.*, 1987). The compounds ethanol, citrate, methanol, butane diol, and glucose can induce phosphorus release only under strict anaerobic conditions. The third group of compounds— butyrate, lactate, and succinate—are found to induce phosphorus release under anoxic conditions from some sludges but not from others. Substrates that are readily taken up by organisms or are high in concentration favor phosphorus removal (Toerien *et al.*, 1990).

The effect of nitrate on phosphorus release depends on substrate composition and the relative amounts of nitrate and substrate. Gerber *et al.* (1987) reported that phosphorus release started only after nitrate was depleted when the substrate was ethanol. However, nitrate did not appear to adversely affect phosphorus release when acetate was the substrate. When substrates that induce phosphorus release only under anaerobic conditions are present, nitrate prevents phosphorus release.

Torien *et al.* (1990) reported that sulfide concentrations of 25 mg/L or higher might be detrimental to phosphorus release. Phosphorus uptake appears to occur only under aerobic conditions.

TRACE ELEMENT REMOVAL

Removal of trace elements by biological processes occurs through biological uptake and adsorption.

SUMMARY

Nutrients essential for growth of organisms include carbon, oxygen, nitrogen, hydrogen, phosphorus, sulfur, calcium, magnesium, potassium, iron, and trace elements. Sodium, chloride, and growth factors are required by some, but not all, organisms. Carbon, oxygen, nitrogen, hydrogen, phosphorus, and sulfur are the main constituents of the cell. The other essential nutrients serve various functions, including acting as enzyme activators, electron transporters, and regulators of osmotic pressure.

Nutrients are required in proper proportions for efficient treatment of wastewater. When nutrients are deficient, sludge settleability and effluent quality in treatment plants deteriorate, leading to process failure. The existence of nutrient deficiency can be determined from influent analyses or cell composition. If nutrient deficiency is confirmed, nutrients can be added in proper ratios to correct the problem. If certain nutrients are in excess, they cannot be removed effectively in treatment plants and can cause deterioration of the quality of receiving waters. Harmful effects of excessive nutrients in receiving waters include eutrophication, fish kills, and making water unfit for various beneficial uses such as potable water supply and recreation. When nutrients are in excess, they can be removed in various ways, including physical, chemical, and biological processes.

REFERENCES

Alroy, Y., and Tannenbaum, S.R. (1973) The Influence of Environmental Conditions on the Macromolecular Composition of *Candida utilis*. *Biotechnol. Bioeng.*, **15**, 239.

American Public Health Association (1989) *Standard Methods for the Examination of Water and Wastewater.* 17th Ed., Washington, D.C.

Becker, W.M. (1977) *Energy and the Living Cell, An Introduction to Bioenergetics.* J.B. Lippincott Co., Philadelphia, Pa.

Brock, T.D., and Madigan, M.T. (1988) *Biology of Microorganisms.* 5th Ed., Prentice Hall, Inc., Englewood Cliffs, N.J.

Broderick, T.A., and Sherrard, J.H. (1985) Treatment of nutrient deficient wastewaters. *J. Water Pollut. Control Fed.*, **57**, 1178.

Brown, C.M., and Rose, A.H. (1969) Effects of Temperature on Composition and Cell Volume of *Candida utilis*. *J. Bacteriol.*, **97**, 261.

Burkhead, C.E., and Waddell, S.L. (1969) Composition Studies of Activated Sludges. *Proc. 24th Ind. Waste Conf., Purdue Univ.*, **135**, 576, West Lafayette, Ind.

Emerson, S.L., and Sherrard, J.H. (1983) Aerobic biological treatment of nitrogen-deficient wastewaters. *J. Water Pollut. Control. Fed.*, **55**, 467.

Genetelli, E.J. (1967) DNA and Nitrogen relationships in bulking activated sludge. *J. Water Pollut. Control Fed.*, **39**, 2, R31.

Gerardi, M.H. (1991) An Operator's Guide to the Microbiological Role of Nitrogen and Phosphorus in the Activated Sludge Process. Unpublished paper.

Gerber, A., *et al.* (1987) Interactions Between Phosphate, Nitrate and Organic Substrate in Biological Nutrient Removal Processes. *Water Sci. Technol.*, **19**, 183.

Gottschalk, G. (1986) *Bacterial Metabolism.* 2nd Ed., Springer-Verlag, New York, N.Y.

Grady, C.P.L., Jr., and Lim, H.C. (1980) *Biological Wastewater Treatment, Theory and Applications.* Marcel Dekker, Inc., New York, N.Y.

Grau, P. (1991) Criteria for Nutrient-Balanced Operation of Activated Sludge Process. *Water Sci. Technol.*, **24**, 251.

Helmers, E.N., *et al.* (1951) Nutritional requirements in the biological stabilization of industrial wastes, II. Treatment with domestic sewage. *Sew. Ind. Wastes*, **23**, 884.

Helmers, E.N., *et al.* (1952) Nutritional requirements in the biological stabilization of industrial wastes, III. Treatment with supplementary nutrients. *Sew. Ind. Wastes*, **24**, 496.

Herbert, D. (1961) The Chemical Composition of Micro-Organisms as a Function of Their Environment. In *Microbial Reaction to Environment.*

The Soc. for Gen. Microbiol., The University Press, Cambridge, London, Eng.

Hodson, P.H. (1973) The Role of Phosphorus in Bacteria and Viruses. In *Environmental Phosphorus Handbook*. E.J. Griffith *et al.* (Eds.), John Wiley & Sons, New York, N.Y.

Hoover, S.R., and Porges, N. (1952) Assimilation of dairy wastes by activated sludge, II. The equation of synthesis and rate of oxygen utilization. *Sew. Ind. Wastes*, **24**, 306.

Jenkins, D., *et al.* (1984) *Manual on the Causes and Control of Activated Sludge Bulking and Foaming*. Rep. Prepared for the Water Res. Comm., Rep. S.A.

Luria, S.E. (1960) Bacterial Protoplasm—Composition and Organization. In *The Bacteria, Volume 1: Structure*. J.C. Gunsales and R.Y. Stainer (Eds.), Academic Press, New York, N.Y.

McCarty, P.L. (1970) Phosphorus and Nitrogen Removal by Biological Systems. *Proc. Wastewater Reclamation and Reuse Workshop*, Lake Tahoe, Calif.

Metcalf & Eddy, Inc. (1991) *Wastewater Engineering—Treatment, Disposal, and Reuse*. 3rd Ed., McGraw-Hill, Inc., New York, N.Y.

Neidhardt, C.F., *et al.* (1990) *Physiology of the Bacterial Cell, A Molecular Approach*. Sinauer Associates, Inc., Sunderland, Mass.

Nowak, G. (1986) Effects of feed pattern and dissolved oxygen on growth of filamentous bacteria. *J. Water Pollut. Control Fed.*, **58**, 978.

Ostrander, S.J. (1992) A Non-Conventional Solution to an Old Problem. *Oper. Forum*, **9**, 10.

Oswald, W.J. (1963) Fundamental Factors in Stabilization Pond Design. In *Advances in Biological Waste Treatment*. W.W. Eckenfelder, Jr., and B.J. McCabe (Eds.), The Macmillan Co., New York, N.Y.

Pirt, S.J. (1975) *Principles of Microbe and Cell Cultivation*. Blackwell, Oxford, Eng.

Pitter, P., and Chudoba, J. (1990) *Biodegradability of Organic Substances in the Aquatic Environment*. CRC Press, Inc., Boca Raton, Fla.

Rickard, M.D., and Gaudy, A.F., Jr. (1968) Effect of mixing energy on sludge yield and cell composition. *J. Water Pollut. Control Fed.*, **40**, 2, R129.

Sawyer, C.N. (1956) Bacterial Nutrition and Synthesis. In *Biological Treatment of Sewage and Industrial Wastes, Volume I: Aerobic Oxidation*. B.J. McCabe and W.W. Eckenfelder, Jr. (Eds.), Reinhold Publishing Corporation, New York, N.Y.

Sawyer, C.N., and McCarty, P.L. (1978) *Chemistry For Environmental Engineering*. 3rd Ed., McGraw-Hill, Inc., New York, N.Y.

Sezgin, M., *et al.* (1988) Isolation and Identification of Actinomycetes Present in Activated Sludge Scum. *Water Sci. Technol.*, **20**, 257.

Sherrard, J.H., and Schroeder, E.D. (1976) Stoichiometry of industrial biological wastewater treatment. *J. Water Pollut. Control Fed.*, **48**, 742.

Snoeyink, V.L., and Jenkins, D. (1980) *Water Chemistry*. John Wiley & Sons, New York, N.Y.

Speece, R.E., and McCarty, P.L. (1964) Nutrient Requirements and Biological Solids Accumulation in Anaerobic Digestion. In *Advances in Water Pollution Research*. Volume 2, W.W. Eckenfelder, Jr. (Ed.), The Macmillan Co., New York, N.Y.

Stanier, R.Y., *et al.* (1986) *The Microbial World*. 4th Ed., Prentice-Hall, Inc., Englewood Cliffs, N.J.

Starkey, J.E., and Karr, P.R. (1984) Effect of low dissolved oxygen concentration on effluent turbidity. *J. Water Pollut. Control Fed.*, **56**, 837.

Stumm, W., and Tenney, M.W. (1964) Waste Treatment for the Control of Heterotrophic and Autotrophic Activity in Receiving Waters. *Proc. 12th Munic. Ind. Waste Conf.*, Raleigh, N.C.

Symons, J.M., and McKinney, R.E. (1958) The biochemistry of nitrogen in the synthesis of activated sludge. *Sew. Ind. Wastes*, **30**, 7.

Tempest, D.W., and Hunter, J.R. (1965) The Influence of Temperature and pH Value on the Macromolecular Composition of Magnesium-limited and Glycerol-limited *Aerobacter aerogenes* Growing in a Chemostat. *J. Gen. Microbiol.*, **41**, 267.

Toerien, D.F., *et al.* (1990) Enhanced Biological Phosphorus Removal in Activated Sludge Systems. In *Advances in Microbial Ecology, Volume 11*. K.C. Marshall (Ed.), Plenum Press, New York, N.Y.

U.S. EPA (1985) *Design Manual, Odor and Corrosion Control in Sanitary Sewerage Systems and Treatment Plants*. Office of Res. and Dev., Cincinnati, Ohio.

U.S. EPA (1975) *Process Design Manual for Nitrogen Control*. Office of Technol. Transfer, Washington, D.C.

U.S. EPA (1976) *Process Design Manual for Phosphorus Removal*. Office of Technol. Transfer, Washington, D.C.

U.S. EPA (1974) *Process Design Manual for Sulfide Control in Sanitary Sewerage Systems*. Office of Technol. Transfer, Washington, D.C.

Varma, M.M., and DiGiano, F. (1968) Kinetics of oxygen uptake by dead algae. *J. Water Pollut. Control Fed.*, **40**, 613.

Water Pollution Control Federation (1987) *Activated Sludge*. Manual of Practice No. OM-9, Alexandria, Va.

Water Pollution Control Federation (1990) *Wastewater Biology: The Microlife*. Special Publication, Alexandria, Va.

Weddle, C.L., and Jenkins, D. (1971) The Viability and Activity of Activated Sludge. *Water Res.*, **5**, 621.

Wood, D.K., and Tchobanoglous, G. (1975) Trace elements in biological waste treatment. *J. Water Pollut. Control Fed.*, **47**, 1933.

Chapter 4
Heterotrophic and Autotrophic Bacteria

WHAT ARE HETEROTROPHIC AND AUTOTROPHIC BACTERIA?

All living organisms can be grouped into broad categories according to how they derive carbon and energy from the environment. Organisms that use carbon dioxide as their carbon source for growth are called *autotrophs*. Plants, algae, and many different kinds of bacteria are autotrophs. All other living organisms are *heterotrophs*, which derive the carbon they need for growth from organic compounds.

Living organisms can further be divided into categories based on the way they obtain energy. All plants and algae and a few bacterial species derive their energy from photosynthesis and are called *phototrophs*. All other organisms derive energy from the oxidation of chemical substances and are called *chemotrophs*. *Chemoorganotrophs* derive energy from the oxidation of organic compounds, and all higher organisms except plants belong to this category. Most heterotrophic bacteria are chemoorganotrophs.

There are a few special groups of bacteria, called *chemolithotrophs* that derive their energy from the oxidation of inorganic compounds such as ammonia or hydrogen sulfide. Most chemolithotrophs obtain their carbon from carbon dioxide and are commonly referred to as *chemoautotrophs*. Nitrifying bacteria are chemoautotrophs. Chemoautotrophs play a crucial role in the recycling of elements such as nitrogen and sulfur. Without their activity, these elements would become locked up in dead organisms and eventually become unavailable for use as nutrients.

The terms that describe the carbon and energy source typically are combined to characterize the major physiological groups of living organisms. *Photoautotrophs* (including plants, algae, and some bacteria) obtain energy from sunlight and carbon from carbon dioxide. They can grow in a low-nutrient environment as long as carbon dioxide; sunlight; and sources of nitrogen, sulfur, and trace nutrients are available.

A few bacterial species are *photoheterotrophs*. They use sunlight for energy and organic compounds for their carbon nutrient and generally are not considered important in wastewater treatment. Organisms that obtain carbon and energy from organic compounds are *chemoheterotrophs*. Most chemoheterotrophs are chemoorganotrophs, which play a key role in waste treatment processes such as activated sludge, tricking filters, rotating biological contactors, and anaerobic digesters by growing on and removing organic carbon compounds from the water.

METABOLISM OF HETEROTROPHS AND AUTOTROPHS

The term *metabolism* refers to all of the biochemical reactions that occur in cells. Catalyzed by enzymes, metabolic reactions are the mechanisms by which cells derive energy and convert nutrients to their own biomass. There are three basic similarities in the metabolism of all living things, whether heterotrophs or autotrophs: (1) all organisms must convert their carbon source to their own cell material; (2) they must have a source of *reducing power*; and (3) they must use their energy source to form adenosine triphosphate (ATP), which in turn can be used as the energy source for biosynthetic reactions in the cell.

Reducing power is the source of hydrogens needed to convert the carbon source to cell material. Organisms growing on organic compounds typically can use the hydrogens from the organic compound itself as the source of reducing power, whereas organisms using carbon dioxide as the carbon source must reduce the carbon dioxide to biomass.

Adenosine triphosphate can be thought of as "energy dollars" provided to the cell by oxidations that can be spent for various energy-requiring activities, such as making new cell material or movement from one place to another.

AEROBIC RESPIRATION. Chemoheterotrophs oxidize organic compounds and derive energy by a series of reactions referred to as metabolic pathways. The reducing power needed to build cell material comes from the oxidation of these same compounds. During the breakdown of the organic nutrients, hydrogens are removed and transferred to a special electron carrier coenzyme called nicotinamide adenine dinucleotide (NAD). The product is the reduced form of NAD, designated as $NADH_2$. The protons from the hydrogens are then transferred to the outside of the cell membrane and the electrons are transferred to a series of membrane-bound proteins and cofactors called the *electron transport system*.

The electrons move from one protein (cytochrome) to the next until they reach the *terminal electron acceptor*, which is oxygen (under aerobic conditions). The resultant charge separation between protons on the outside and electrons on the inside of the membrane creates a miniature battery. The potential energy created by this gradient is used to form ATP. This involves a special enzyme system in the membrane called ATPase, which couples the return of the protons through the cell membrane with ATP formation. Chemoheterotrophs, therefore, derive their carbon, energy, and reducing power from the organic compounds on which they grow.

ANAEROBIC RESPIRATION. Some microorganisms are able to carry out *anaerobic respiration*, in which inorganic compounds such as nitrate or sulfate are used as the terminal electron acceptor in the absence of oxygen.

Denitrification is an example of anaerobic respiration. In this process, nitrate is the terminal electron acceptor and organic compounds such as methanol serve as electron donors. In wastewater treatment processes, the nitrate produced during nitrification can be removed by adding an inexpensive organic compound such as methanol or raw wastewater to a microbial population under anaerobic conditions.

This process is carried out by a variety of heterotrophic bacteria found in activated sludge that are capable of denitrification, including species of *Alcaligenes*, *Achromobacter*, *Bacillus*, *Hyphomicrobium*, *Micrococcus*, and *Pseudomonas*. These same organisms can use oxygen when it is available, which is why anaerobic conditions are required for nitrate removal. A high oxygen concentration inhibits denitrification by inhibiting the enzymes involved in the process (Brock and Madigan, 1991).

Sulfate also can be used as a terminal electron acceptor. In this case, the process is referred to as *dissimilative sulfate reduction* and the sulfate is reduced to hydrogen sulfide. Organic compounds generated under anaerobic conditions, such as lactate and acetate, are used as the source of carbon and reducing power and hydrogen sulfide is the final product.

Production of hydrogen sulfide, a gas detectable in low concentrations, causes odors in wastewater treatment systems. The predominant bacterial genus involved in conversion of sulfate to hydrogen sulfide is *Desulfovibrio*.

FERMENTATION. Some chemoheterotrophs, referred to as anaerobes, are unable to use oxygen or inorganic compounds as the terminal electron acceptor. These organisms must use a broken-down product of their growth substrate as the terminal electron acceptor through a process called *fermentation*. Some organisms can use oxygen when it is available but can carry out fermentation when oxygen is not present. During fermentation, growth substrates like glucose may be broken down to pyruvic acid, which serves as the terminal electron acceptor. The final product is lactic acid, some other acid, or ethanol.

$$
\begin{array}{ccc}
\text{Glucose} & \text{Pyruvic acid} & \text{Lactic acid} \\
C_6H_{12}O_6 \rightarrow & 2CH_3COCOOH \rightarrow & 2CH_3CHOHCOOH
\end{array} \qquad (4.1)
$$

Compared with aerobic processes, when fermentation occurs, relatively little energy is released from the growth substrate. Most of the energy remains in the final product (lactic acid). Under aerobic conditions, more energy is released through oxidation of the substrate to carbon dioxide.

CHEMOLITHOTROPHY. Chemolithotrophs derive energy from the oxidation of inorganic compounds. In *nitrification*, ammonia (NH_3) is oxidized to nitrite by *Nitrosomonas* and nitrite is then oxidized to nitrate by *Nitrobacter*. The nitrifying bacteria require oxygen as the terminal electron acceptor.

Organisms like *Thiobacillus* oxidize hydrogen sulfide to sulfate. Other organisms, such as *Thiothrix*, may be able to oxidize hydrogen sulfide to elemental sulfur, which is stored intracellularly as sulfur granules. Most of these organisms are autotrophs, but a few can incorporate small amounts of organic compounds. Oxygen typically is used as the terminal electron acceptor.

Chemolithotrophs typically derive less energy from the oxidation of inorganic compounds than heterotrophs derive from the oxidation of organic compounds. This accounts for the dramatic difference in growth rate and growth yield between chemoheterotrophs and nitrifying bacteria.

INCORPORATION OF CARBON DIOXIDE. Autotrophs must reduce carbon dioxide to cell material. They incorporate the carbon dioxide by a series of biochemical reactions. Because cell material is more reduced (has more hydrogens or electrons) than carbon dioxide, these organisms must have an external source of reducing power from which to obtain hydrogens or electrons. For plants, algae and cyanobacteria, the reducing power is generated from water during photosynthesis. For chemoautotrophs and phototrophic bacteria other than cyanobacteria, the reducing power normally comes from the oxidation of an inorganic compound such as hydrogen sulfide.

*R*ELATION OF HETEROTROPHIC AND AUTOTROPHIC BACTERIA TO OXYGEN

Bacteria fall into five categories according to their oxygen requirements (Stanier *et al.*, 1986). For growth, *strict aerobes* require oxygen as the terminal electron acceptor. A major portion of the microbial populations in aerated treatment systems are strict aerobes. *Facultative anaerobes* can grow in the presence or absence of oxygen if an appropriate carbon source is available. Many organisms found in the intestinal tract are facultative anaerobes. To grow anaerobically, these organisms require a fermentable substrate or an inorganic compound that can serve as a terminal electron acceptor. *Microaerophiles* require oxygen, but at reduced oxygen tension. These organisms often live in a microenvironment near strict aerobes where lower levels of oxygen are available.

Floc particles in activated sludge represent an oxygen gradient from the outside to the inside of the particle. Rotating biological contactors also represent an environment in which there is a gradient from high to low oxygen tension. *Aerotolerant anaerobes* are organisms that can grow in the presence of oxygen, but are unable to use it for metabolism, typically because of the lack of an electron transport system. *Strict anaerobes* are unable to grow in the presence of oxygen. They carry out fermentation or use an inorganic compound as a terminal electron acceptor. These organisms play a key role in anaerobic digesters.

Oxygen can be harmful to organisms because of the formation of hydrogen peroxide and other toxic compounds by the microbes themselves when oxygen is present. Organisms that can grow in the presence of oxygen must have enzymes that protect them from the toxic effects of oxygen. Strict anaerobes lack these protective enzymes.

Chemoheterotrophic bacteria are represented in all five categories of relationships to oxygen. Chemoautotrophs involved in nitrification are strict aerobes, whereas those involved in the oxidation of sulfur compounds are either strict aerobes or are able to carry out anaerobic respiration.

ROLE OF HETEROTROPHS IN AEROBIC WASTEWATER TREATMENT PROCESSES

DEGRADATION OF ORGANIC COMPOUNDS. The major role of heterotrophic bacteria in wastewater treatment is the removal of soluble and insoluble organic compounds. Most of the active microbial biomass involved in aerobic waste treatment processes consists of heterotrophic organisms. Activated sludge mixed liquor typically contains 10^7 to 10^8 viable heterotrophic bacteria per millilitre (Pike and Curds, 1971). Total numbers depend on the type of wastewater, the retention time, and other physical conditions of the treatment process. Comparison of cell numbers with microbiological culturing techniques indicates that the percent of cells that are viable also depends on environmental conditions but typically is between 1 and 10%.

Heterotrophic bacteria can degrade most, if not all, naturally occurring substances. Some of these materials, such as lignin and humus, are degraded slowly. The specific kinds or organisms present in a waste treatment process will depend on the composition of the wastewater. Aerobic and facultative bacteria predominate in aerobic treatment processes, whereas facultatives and strict anaerobes predominate in anaerobic treatment processes.

MACROMOLECULES. Macromolecules—including proteins, polysaccharides, lipids and nucleic acids—are not transported to the cells of heterotrophs before digestion, but must be broken down into their subunits by enzymes (referred to as *exoenzymes*) that are excreted to the medium (Table 4.1). Only a few bacterial species possess the ability to produce one or more of these exoenzymes. Consequently, these organisms are important in processes treating wastewater containing macromolecules as major constituents.

LOW MOLECULAR WEIGHT COMPOUNDS. The large variety of subunits resulting from the breakdown of macromolecules serve as nutrients

Table 4.1 Subunits formed from the breakdown of macromolecules by heterotrophic bacteria.

Macromolecule	Subunit	Enzyme	Representative organism involved in degradation
Protein	Amino acids	Protease	*Bacillus* *Nocardia* Actinomycetes *Clostridium*
Polysaccharide	Monosaccharides	Amylase Cellulase	*Bacteroides* *Clostridium*
Lipid	Fatty acids	Lipase	*Mycobacterium*
Nucleic acids	Nucleotides	Nucleases	*Streptomyces*
Bacterial cell wall	Amino sugars	Glycanase	*Myxococcus*

for a broad variety of different heterotrophic bacteria. These small molecules are transported to the cell and used for energy and growth. Growth of heterotrophic organisms on organic compounds (such as glucose, acetate, and methanol) depends on a variety of physical conditions and the availability of trace elements.

Physical conditions that influence growth on organic molecules include biomass density, nutrient concentration, aeration, pH, and temperature. Many heterotrophs can grow on a single organic compound as its sole carbon and energy source, whereas others require growth factors such as amino acids and vitamins. When complex wastes such as wastewater are treated, these growth factors are either present in the wastewater or obtained as the result of metabolic activity by other organisms. Some species of *Pseudomonas* can grow on as many as 90 different organic compounds, including many hazardous chemicals, as the sole source of carbon and energy (Stanier *et al.*, 1966).

When the composition of the wastewater changes, there can be a corresponding shift of the microbial population to accommodate the new nutrients. The microbial community undergoes ecological succession, and many organisms can regulate their own metabolism to turn on new enzymes and turn off old ones when available nutrients in wastewater change. Organisms commonly found in aerobic biological waste treatment systems are the following:

- *Achromobacter;*
- *Acinetobacter;*
- *Aerobacter;*
- *Arthrobacter;*
- *Azotobacter;*
- *Bacillus;*
- *Beggiatoa;*

- *Comamonas*;
- *Corynebacterium*;
- *Flavobacterium*;
- *Hyphomicrobium*;
- *Microbacterium*;
- *Micrococcus*;
- *Nocardia*;
- *Pseudomonas*;
- *Sarcina*;
- *Sphaerotilus*;
- *Spirillum*;
- *Thiothrix*; and
- *Zoogloea*.

XENOBIOTICS. Some heterotrophs can degrade both naturally occurring substances and synthetic compounds such as herbicides, insecticides, chlorinated hydrocarbons, and various chemicals used in industry. These compounds have molecular structures that differ from those of naturally occurring compounds and are referred to as *xenobiotics* (Atlas and Bartha, 1987).

Some xenobiotics are degraded as the result of *co-metabolism*, where two or more organisms work together to degrade a single compound. One organism carries out part of the degradation, which renders another portion of the molecule susceptible to degradation by the other organisms. Co-metabolism often is enhanced by the addition of readily usable nutrients. For example, the addition of glucose and nitrogen supplements enhanced the degradation of 2,4-dinitrophenol (Hess and Silverstein, 1990). In other cases, one compound induces enzymes, which can degrade a compound that normally is resistant.

ADAPTABILITY OF THE MICROBIAL BIOMASS. The active biomass of biological waste treatment processes contains a large number of different bacterial genera. This diversity can be accounted for by the variety of nutrient substrates present to support different organisms in wastewater and by the diversity of microenvironments that exist in most biological waste treatment processes. For example, a floc particle consists of a gradient of nutrient and oxygen availability. Organisms that differ in affinity for nutrients and oxygen tension, therefore, can occupy different habitats in the same treatment process.

A treatment process with a well-established, diverse microbial flora typically is somewhat resistant to short-term, minor fluctuations in environmental conditions. When different physical or nutritional conditions such as pH, oxygen concentration, or temperature occur, a significant change in the microbial flora can result. Numbers of organisms of a given species may continually change in response to these conditions, but the end result is continuous, efficient waste removal.

FLOC FORMATION, SLUDGE BULKING, AND FOAMING. In addition to degrading organic wastes, another role of heterotrophs is the formation of floc particles. Floc formation is enhanced by the activities of "zoogloeal" bacteria, which produce extracellular polymers (capsules, slime, glycocalyx) and cause individual bacterial cells to aggregate to particles large enough to settle in a clarifier. Bacteria incapable of extracellular polymer formation adhere to the polymer formed by zoogloeal organisms.

Floc formation is as important a goal of treatment plant operation as is the removal of organic compounds. Problems arising when floc formation does not occur properly include dispersed growth, pin-point floc, rising sludge, and slime bulking. When filamentous organisms predominate, two operational problems can arise: bulking and foaming.

Bulking sludge is defined as one with a high sludge volume index. Many filament bacteria have been implicated in sludge bulking. Their dominance in wastewater treatment systems is thought to stem from specific environmental conditions, including organic loading, nutrient limitation, and plant operating parameters such as sludge age, dissolved oxygen, or intrusion of toxic chemicals (Table 4.2). The various filamentous organisms implicated in bulking, and methods of controlling them, have been described in detail (Strom and Jenkins, 1984; U.S. EPA, 1987; and WPCF, 1990). Control of filamentous bulking depends on the causative agent and typically involves changes in nutrient addition, aeration, or sludge age.

Extensive foam and scum formation can interfere with treatment plant operation. Just as with bulking, specific heterotrophic organisms (primarily of the genus *Nocardia*) are associated with this problem. Control methods for foaming and scum formation have been described (Gerardi, 1986; Lemmer, 1986; Sezgin and Karr, 1986; and WPCF, 1990).

Table 4.2   Dominant filament types as indicators of conditions causing activated sludge bulking.

Suggested causative conditions	Indicative filament types
Low dissolved oxygen	Type 1701, *Sphaerotilus natans*, *Haliscomenobacter hydrossis*
Low food-to-microorganism ratio	*Microthrix parvicella, H. hydrossis, Nocardia* sp., Types 021N, 0041, 0675, 0092, 0581, 0961, 0803
Septic wastewater/sulfide	*Thiothrix* sp., *Beggiatoa*, and Type 021N
Low nutrients (phosphorus or nitrogen)	*Thiothrix* sp., *S. natans*, Type 021N, and possibly *H. hydrossis* and Types 0041 and 0675
Low pH (< 6.5)	Fungi

EFFECTS OF ENVIRONMENTAL CONDITIONS ON HETEROTROPHIC ACTIVITY

pH. Heterotrophic bacteria typically grow best at a pH near 7.0, with a minimum pH of approximately 5.5 and a maximum between 8.5 and 9.5 (Brock and Madigan, 1991). However, each different species is characterized by its own minimum, optimum, and maximum pH values. Some bacteria, such as those that produce acid, can grow at pH values as low as 4.0 to 4.5, whereas those that create an alkaline environment (urea oxidizers) can withstand pH values greater than 9.5.

TEMPERATURE. Seasonal variations in temperature can markedly influence the makeup of microbial communities. Just as with pH, each species is characterized by a minimum, optimum, and maximum temperature that will support growth. *Psychrophiles* grow within the range of 0 to 20°C (32 to 70°F), with an optimum of 10 to 15°C (50 to 60°F). *Mesophiles*, which comprise most of the species commonly found in wastewater treatment processes, grow within the range of 10 to 45°C (50 to 115°F), with an optimum of approximately 30 to 35°C (85 to 95°F). *Thermophiles*, found in compost piles and other high-temperature environments, grow within the range of 40 to 75°C (105 to 165°F), with an optimum growth rate at 55 to 65°C (130 to 150°F). A few species of heterotrophic bacteria, classified as *extreme thermophiles*, can grow at temperatures higher than 100°C (212°F). These organisms live in highly specialized environments, such as geothermal vents in the ocean floor, and have not been implicated in wastewater treatment processes.

ROLE OF HETEROTROPHS IN ANAEROBIC WASTEWATER TREATMENT PROCESSES

Chemoheterotrophs are responsible for degradation of wastes in anaerobic digesters used for stabilization of primary and secondary sludges from industrial and municipal wastewater treatment and for treatment of high-strength industrial wastewater. The organisms involved are either strict anaerobes or facultative anaerobes. The density of heterotrophs in anaerobic digesters may be as high as 10^9 to 10^{10} cells/mL.

In an anaerobic digester, complex organic macromolecules, including microbial biomass, are *depolymerized* (broken down to smaller molecules) and then metabolized primarily to fatty acids, carbon dioxide, and hydrogen gas.

The hydrogen gas may then be used in methane formation. Just as in aerobic waste treatment, a large number of different organisms participate in these processes, including hydrolyzing organisms, acid-forming organisms and *methanogens* (methane-producing organisms). Methane is generated by the direct reduction of methyl groups or by the reduction of carbon dioxide to methane, using hydrogen gas as the reducing agent. Some of the heterotrophic organisms involved in anaerobic digestion are listed in Table 4.3.

Table 4.3   Heterotrophic bacteria involved in anaerobic digestion.

Organism	Role
Clostridium	Cellulose, protein, and nucleic acid degradation
Bacteroides	Cellulose, protein, starch degradation
Ruminococcus	Cellulose degradation
Bacillus	Cellulose degradation
Succinimonas	Starch degradation
Streptococcus	Starch, succinate degradation
Anaerovibrio	Fat degradation
Succinovibrio	Glucose fermentation
Eubacterium	Glucose fermentation
Lactobacillus	Glucose fermentation
Veillonella	Lactate degradation
Methanobacterium	Methane production
Methanobrevibacter	Methane production
Methanococcus	Methane production
Methanosarcina	Methane production

CONTROL OF ANAEROBIC PROCESSES AND FACTORS PROMOTING OPTIMUM GROWTH AND WASTEWATER STABILIZATION. Standard operation of an anaerobic digester requires careful control of pH, temperature, alkalinity, nutrient levels, retention time (30 to 60 days), and organic loading (0.48 to 1.6 kg volatile solids/m^3 of digester volume per day [0.03 to 0.1 lb volatile solids/cu ft of digester volume per day]). Reactors typically are maintained at 35 to 37°C (95 to 100°F) with a pH near 7.0. Extremes of pH, temperature, or the influx of heavy metals or other toxic substances can be disastrous. Long retention times often are maintained because of slow growth of methanogenic bacteria. Digester upset and failure often is related to an imbalance in conversion of fatty acids to methane.

A high-rate system involving two separate stages for fermentation and solids separation was developed as an improvement of the standard method (Metcalf & Eddy, 1979). The retention time for this system is 10 to 15 days, and the organic loading is 1.6 to 2.24 kg volatile solids/m^3 (0.1 to 0.14 lb volatile solids/cu ft) of digester volume per day.

When the pH falls below 6.0, problems are likely to arise. Consequently, monitoring of pH is important.

ROLE OF AUTOTROPHS IN WASTEWATER TREATMENT PROCESSES

Autotrophic bacteria play a key role in the cycling of elements in the natural environment and in wastewater treatment processes. These organisms derive carbon from carbon dioxide and, therefore, do not contribute directly to biochemical oxygen demand (BOD) removal. They derive energy from the oxidation of inorganic compounds present in the water or generated from the decomposition of nutrients by heterotrophs. Examples of substrates for autotrophs include ammonia, nitrite, hydrogen sulfide, elemental sulfur, and hydrogen gas. Most of the chemoautotrophs are strict aerobes, whereas all photoautotrophic bacteria (except cyanobacteria) are anaerobes. Table 4.4 lists the major groups of chemoautotrophic bacteria and their energy sources.

Because less energy is derived from the oxidation of inorganic compounds than from organic compounds, autotrophs typically grow more slowly than heterotrophs and the cell yield is much lower. The doubling time for heterotrophs may be 0.5 to 2 hours, whereas nitrifying bacteria require a minimum of 24 to 48 hours to double in numbers. Consequently, autotrophs do not compete well with heterotrophs for trace nutrients or oxygen. Standard methods for enumerating bacteria, such as plating on agar media, are often more difficult or impossible for autotrophs because the yield is so low that colonies can barely be detected. Turbidity of liquid cultures is barely visible for most chemoautotrophs.

Table 4.4   Major groups of chemoautotrophic bacteria.

Group	Genera	Substrate oxidized and product formed
Nitrifying bacteria	*Nitrosomonas*	Ammonia → nitrite
	Nitrobacter	Nitrite → nitrate
Sulfur bacteria	*Thiobacillus*	Sulfide → sulfate
	Beggiatoa	
	Thiothrix	
Iron bacteria	*Sphaerotilus*	Ferrous → ferric
	Leptothrix	Ferrous → ferric
	Crenothrix	Ferrous → ferric
	Gallionella	Ferrous → ferric

Nitrification

Under aerobic conditions, ammonia is removed biologically by a two-step process in which the ammonia is oxidized to nitrite, and the nitrite is oxidized to nitrate (U.S. EPA, 1975). The two steps taken together are referred to as nitrification:

$$NH_3 + \tfrac{3}{2} O_2 \rightarrow NO_2^- + H^+ + H_2O \qquad (4.2)$$

$$NO_2^- + \tfrac{1}{2} O_2 \rightarrow NO_3^- \qquad (4.3)$$

The overall reaction for nitrification is

$$NH_3 + 2 O_2 \rightarrow NO_3^- + H^+ + H_2O \qquad (4.4)$$

Nitrifying bacteria use carbon dioxide and bicarbonate alkalinity as carbon sources and are unable to use significant amounts of organic compounds for growth. As can be seen from the equations above, nitrification results in acid formation. Because large quantities may be produced, a decrease in pH can occur in poorly buffered environments.

Although *Nitrosomonas* is the organism most frequently associated with the conversion of ammonia to nitrite, *Nitrosococcus*, *Nitrosospira*, and *Nitrosolobus* are also capable of carrying out this reaction. *Nitrobacter* is most commonly associated with oxidation of nitrite to nitrate. Other nitrite oxidizers include *Nitrococcus*, *Nitrocystis*, and *Nitrospira*. Most of these genera are common inhabitants of soil and water and grow as single cells or in clusters of cells held together by a diffuse slime layer. The nitrifying bacteria require no growth factors and are capable of growing on completely inorganic media.

Control of Nitrification

TEMPERATURE. The rate of growth on nitrifying bacteria varies considerably with wastewater temperature. The optimum temperature range in suspended growth systems, such as activated sludge, is approximately 25 to 30°C (75 to 85°F). Temperature affects the growth rate of nitrifying bacteria and, therefore, the numbers of nitrifying bacteria. Temperature also affects the rate of ammonia conversion to nitrite and nitrate. Nitrification is greatly reduced below 18°C (65°F). Because the inhibitory effect of decreased temperature is greater on nitrate formers than on nitrite formers, a buildup of nitrites can occur during depressed temperature, which may inhibit conversion of ammonia to nitrite. An increase in nitrite concentration may also be toxic to other bacteria.

---

pH. The optimum pH range for nitrification is on the alkaline side of neutrality (between 7.5 and 8.5). A higher pH tends to maintain un-ionized ammonia (NH_3), the form used by nitrifiers. Although nitrifying bacteria are pH sensitive, they have the ability to acclimate to pH values outside the optimum range. The adaptation may require several weeks of gradual increase or decrease of pH, and steady conditions of pH must be maintained when operating outside the optimum range.

OXYGEN. Nitrification places a significant oxygen demand on the activated sludge process. Conversion of ammonia requires approximately 4 kg (8 lb) of oxygen per kg (lb) of ammonia converted. Nitrification can occur in activated sludge with a dissolved oxygen concentration of 0.5 mg/L (ppm), but it is only minimal below 1.0 mg/L (ppm). Increasing the dissolved oxygen concentration enhances the nitrification process.

SENSITIVITY TO HEAVY METALS AND OTHER TOXIC SUBSTANCES. Nitrifying bacteria are sensitive to even low concentrations of heavy metals and other substances found in industrial wastewater. Also, nitrifying bacteria are sensitive to high concentrations of ammonia and nitrite. Therefore, excessive ammonification of organic matter can be inhibitory to nitrification. When ammonia is converted to nitrite at a faster rate than the nitrite is converted to nitrate, the entire process can be inhibited. Inhibition of nitrification through high ammonia levels can be overcome by reducing the daily nitrogen loading rate (U.S. EPA, 1975).

OPERATIONAL FACTORS. High nitrification efficiencies can be attained by maintaining acceptable pH, temperature, and aeration (greater than 1.0 mg/L [ppm] dissolved oxygen), increased contact time (higher levels of solids return from the secondary clarifier to the mixed liquor basin), and increased mean cell residence time (MCRT). Increasing MCRT is often an important factor in improving nitrification.

NITRIFICATION AND DENITRIFICATION IN THE SAME FLOC. Denitrification may occur in mixed liquor if the oxygen concentration is low enough (less than 1.0 mg/L [ppm]). If an anoxic condition develops in the sludge floc, as occurs when oxygen is used at the same rate it is supplied, the inside of floc particles will be anaerobic. This requires that the floc particles be relatively large. Simultaneous nitrification and denitrification can be attained through intermittent aeration in a single basin, but is operationally difficult to achieve.

OXIDATION OF SULFUR AND IRON COMPOUNDS

COLORLESS SULFUR BACTERIA. Colorless sulfur bacteria are aerobes that obtain energy by oxidizing sulfur compounds. The major genera of bacteria important in wastewater treatment are *Thiobacillus*, *Thiothrix*, and *Beggiatoa*.

Beggiatoa and some of the thiobacilli, such as *Thiobacillus novellus* and *Thiobacillus intermedius*, are *mixotrophs*, which means they are capable of growing as autotrophs or heterotrophs, depending on the nutrients available. Although some of the thiobacilli are restricted to environments of near-neutral pH, many species, including *Thiobacillus thiooxydans* and *Thiobacillus ferrooxydans*, can grow well in acid environments (below pH 2.0). Most thiobacilli can oxidize sulfur and sulfur compounds such as hydrogen sulfide. The final product of oxidation is sulfuric acid, which can result in a low pH.

IRON BACTERIA. Iron bacteria are commonly found in water that contains iron. Growth of these organisms is characterized by encrusted iron oxides or manganese buildup on the surfaces of the cells. Iron is deposited in the capsular material or on the sheath (*Leptothrix*) or stalk (*Gallionella*) that surrounds the cells. Genera of iron-oxidizing filamentous bacteria include *Sphaerotilus*, *Leptothrix*, *Crenothrix*, and *Gallionella*. These organisms appear to use ferrous iron as an energy source, producing ferric ions that form insoluble ferric hydroxide. Because ferrous is rapidly converted to ferric in the presence of oxygen, growth of iron-oxidizing bacteria requires reduced oxygen tension.

Many iron-oxidizing bacteria have been implicated in sludge bulking. They are also responsible for clogging of pipes and water-handling equipment when they form thick slime layers on surfaces.

INTERACTIONS BETWEEN AUTOTROPHS AND HETEROTROPHS

In the natural environment and in wastewater treatment processes, organisms do not function in isolation but work together to carry out waste removal. Some organisms carry out the degradation of macromolecules (such as cellulose and proteins) whereas others degrade products of their metabolic activities (such as glucose and amino acids). Some organisms must work together to degrade a single substance (co-metabolism). Therefore, successful operation of a treatment plant involves maintaining conditions for the proper interaction of heterotrophs and autotrophs. In some cases, interactions between

---

different organisms interfere with proper waste treatment. Various microbial interactions are best understood in the context of biocycling of the elements.

THE CARBON CYCLE. Carbon is cycled between carbon dioxide and organic compounds (Figure 4.1). Plants and photosynthetic organisms are responsible for autotrophic conversion of atmospheric carbon dioxide to biomass. Carbon dioxide is returned to the atmosphere by heterotrophs carrying out aerobic respiration and fermentation. Under anaerobic conditions, organic nutrients can never be completely oxidized to carbon dioxide. A portion of the original substrate must be used as the terminal electron acceptor, which through fermentation results in formation of smaller organic compounds such as acetate, lactate, ethanol, and methane.

THE NITROGEN CYCLE. The activities of organisms are essential in maintaining the operation of the nitrogen cycle (Figure 4.2). Ammonia is first released from organic compounds, such as proteins, by heterotrophs. Although some of this ammonia is immediately re-incorporated by organisms as they grow, the rest is released to the environment. Under aerobic conditions, ammonia can be oxidized to nitrite and then to nitrate by nitrifying bacteria. The nitrate can then be used by plants and other organisms as a source of nitrogen. Under anaerobic conditions, ammonia cannot be oxidized. Only a small amount can be removed by assimilation to growing organisms.

Under anaerobic conditions, nitrate can be converted to nitrogen gas by heterotrophic denitrifying bacteria. This requires the presence of organic compounds to serve as a source of reducing power. Under aerobic conditions, nitrate can be assimilated by living organisms, but cannot be converted to nitrogen gas.

NITROGEN FIXATION. Nitrogen fixation is the only biological mechanism for returning nitrogen from the atmosphere to biological material:

$$N_2 \rightarrow 2\,NH_3 \qquad\qquad (4.5)$$

The conversion of nitrogen gas to ammonia is catalyzed by a special enzyme called nitrogenase. Nitrogen fixation is carried out by free-living bacteria (such as *Azotobacter* and other bacterial species in the soil), photoautotrophic cyanobacteria, and *Rhizobium* (a bacterium living in symbiotic association with the root nodules of leguminous plants). Without the action of the nitrogen-fixing bacteria, nitrogen lost to the atmosphere would be lost and all life eventually would cease. The potential for bacterial nitrogen fixation has not been applied to treatment of low-nitrogen wastes.

THE SULFUR CYCLE. Although required in smaller amounts than nitrogen, sulfur is an essential nutrient. The degradation of municipal wastes by

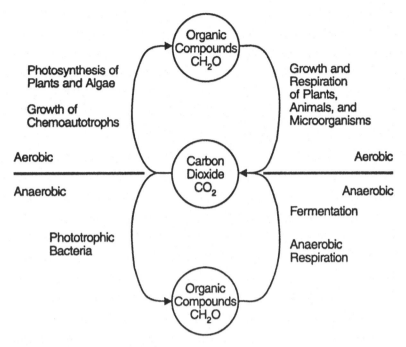

Figure 4.1   The carbon cycle.

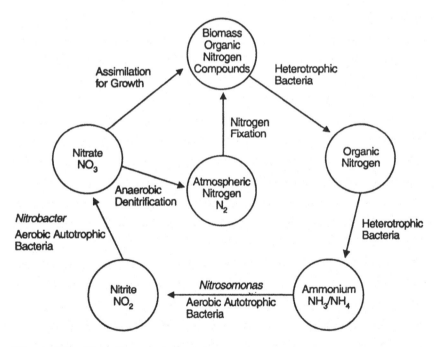

Figure 4.2   The nitrogen cycle.

Heterotrophic and Autotrophic Bacteria   *81*

heterotrophic bacteria results in the release of sulfate (Figure 4.3). Under anaerobic conditions, organic compounds are oxidized and sulfate is reduced to hydrogen sulfide by *Desulfovibrio*. Hydrogen sulfide cannot be further metabolized under anaerobic conditions except by photoautotrophic bacteria that use hydrogen sulfide as an electron donor. As the hydrogen sulfide moves to an aerobic environment, it is used as an energy source by chemoautotrophs, such as *Thiobacillus*, or is spontaneously oxidized. The sulfate can then be incorporated by plants and other higher organisms.

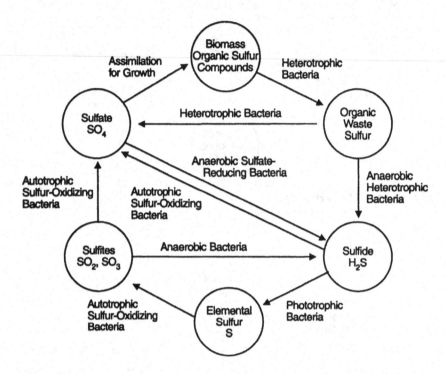

Figure 4.3   The sulfur cycle.

CORROSION. There are different mechanisms of corrosion, including fatigue, stress, fretting, cavitation, filiform, and bacteriological. Bacteriological corrosion involves the combined action of anaerobic sulfate-reducing chemoheterotrophs and aerobic sulfide-oxidizing chemoautotrophs (Bowker and Smith, 1985). When wastewater containing sulfate and organic matter becomes anaerobic, hydrogen sulfide is produced through the metabolic activities of anaerobic sulfate-reducing bacteria. The hydrogen sulfide is released from the water and then dissolves in the condensed moisture on structures above the surface. *Thiobacilli* then oxidize the sulfide, resulting in corrosion of concrete and other structures associated with waste treatment processes.

Understanding the nature of microbial activities involved in corrosion suggests ways to reduce or prevent the problem. These include improving the oxygen balance by air injection and chemical addition to inhibit sulfate reduction, cause precipitation of oxidation products, and control pH (Sims, 1980, and Speece and Orosco, 1970).

PHOTOTROPHIC ORGANISMS

Photosynthesis is carried out by algae and by four groups of bacteria: purple sulfur bacteria, nonsulfur purple bacteria, green bacteria, and cyanobacteria. Plants, algae, and cyanobacteria carry out *oxygenic* photosynthesis, where water is used as an electron donor and oxygen is produced. All of these organisms contain a particular type of chlorophyll referred to as chlorophyll *a*. Purple and green bacteria carry out *anoxygenic* photosynthesis (no oxygen is produced because hydrogen sulfide [H_2S] rather than water [H_2O] is used as the electron donor) and contain a variety of different forms of chlorophyll referred to as bacteriochlorophylls (Brock and Madigan, 1991).

Photosynthetic activity of organisms is important in wastewater treatment involving aerobic and facultative ponds. Although oxygenation and biomass production by algae and cyanobacteria have been considered in the design of aerobic and facultative wastewater treatment ponds, the role and significance of green and purple bacteria has been overlooked. Under appropriate conditions, their growth can result in production of biomass from nutrients not introduced in the influent (light and carbon dioxide), which can contribute significantly to effluent suspended solids.

Algae and cyanobacteria are photoautotrophs. They derive carbon for biosynthesis of cell material from carbon dioxide. The carbon dioxide is incorporated to the cells by a series of biochemical reactions. Light energy is captured in chlorophyll and converted to stored energy as ATP. In algae and cyanobacteria, water is the source of reducing power (hydrogens) needed to convert carbon dioxide (CO_2) to cell biomass.

The purple and green bacteria grow photoautotrophically only under anaerobic conditions. In the presence of oxygen, photosynthetic pigment production is repressed. Although these organisms obtain energy from photosynthesis, they are unable to use water as a reductant. They must, therefore, obtain reducing power for biosynthesis from either a reduced sulfur compound or hydrogen gas. Some purple and green bacteria can grow heterotrophically in the dark, using organic compounds as their source of carbon and energy and oxygen or some other compound as the terminal electron acceptor.

PRINCIPAL GROUPS OF PHOTOSYNTHETIC ORGANISMS

PURPLE SULFUR BACTERIA. Purple sulfur bacteria are restricted to growth in anaerobic zones of lakes, ponds, or sulfur springs where sulfide is provided by sulfate reduction in the sediment below. These organisms have a limited ability to use organic compounds as a source of carbon. Because the chlorophyll pigments in these organisms absorb light at longer wavelengths than those of algae and cyanobacteria, they can inhabit the water column at a lower position and, therefore, do to compete with oxygenic organisms.

NONSULFUR PURPLE BACTERIA. These bacteria have been called non-sulfur purple bacteria because it was originally thought that they could not use sulfur compounds as electron donors. It is now known that sulfide is toxic to these organisms at the concentrations tolerated by purple sulfur bacteria, but that they can use sulfide at lower concentrations. Many nonsulfur purple bacteria are facultative phototrophs that can grow photosynthetically in the presence of light and absence of oxygen. They also can grow chemoheterotrophically in the presence of an organic substrate and oxygen.

GREEN BACTERIA. Green photosynthetic bacteria are strict anaerobes and obligate phototrophs. Although they are capable of assimilation of small organic molecules such as acetate, propionate, and pyruvate, they require the presence of reduced sulfur compounds for growth.

CYANOBACTERIA. Cyanobacteria, formerly known as blue-green algae, are widely distributed in soil and aquatic environments. They carry out photosynthesis and generate oxygen. There are a variety of morphological types, including cocci (round), aggregates of cells (colonies), and filaments. Cyanobacteria are more tolerant of environmental extremes than are algae, and some of them are capable of nitrogen fixation. Some cyanobacteria produce *geosmins*, small molecules with complex structures that release an odor to treated water.

ALGAE. Algae contribute to the oxygen content of water by producing oxygen during photosynthesis. There are several groups of algae based on pigment content and type of storage polymer found in the cells. There are several morphological types, including filaments, cocci, and clusters of cells.

ROLES PERFORMED BY PHOTOSYNTHETIC ORGANISMS

Where large land areas are available, ponds and stabilization basins can be used for wastewater treatment. They are the most common method of treatment in developing countries.

AEROBIC ALGAE PONDS. Algae and cyanobacteria provide the oxygen for BOD removal in aerobic ponds. Organic loading must be low enough to be removed by oxygen produced by algae. Most aerobic ponds are limited in depth to approximately 0.5 m (1.5 ft) to permit light penetration (Benefield and Randall, 1980). As algae are sensitive to a variety of toxic substances, aerobic ponds are limited to treatment of nontoxic wastewater.

Specific types of algae found in aerobic ponds may vary seasonally. Algae often are replaced by cyanobacteria when the temperature exceeds 35°C (95°F). Aerobic ponds should be mixed periodically to prevent thermal stratification. Without primary treatment, solids will settle to the bottom and create an anaerobic zone.

FACULTATIVE PONDS. Facultative ponds are used frequently by small communities. The depth of a facultative pond is from 1 to 2.5 m (3 to 8 ft). When thermal stratification occurs, the upper layer is aerobic during the day and the bottom layer is anaerobic. Waste solids at the bottom of the pond are anaerobically digested, releasing methane and other anaerobic products to the aerobic layer. With BOD loadings of approximately 20 to 55 kg/hd (20 to 50 lb/ac/d) and retention times of 7 to 50 days, 70 to 90% BOD removal can be expected (Benefield and Randall, 1980). However, effluent suspended solids because of algae may be 100 to 350 mg/L (ppm).

ANAEROBIC PONDS. A pond will operate anaerobically if the influent BOD exceeds the oxygen production from photosynthesis. Anaerobic ponds are used primarily for pretreatment and are particularly suited for high-temperature or high-strength wastewater. Use of anaerobic pretreatment results in reduction in accumulated sludge in subsequent basins.

Anaerobic ponds are constructed to various depths (1 to 8 m [3 to 25 ft]). Deep ponds offer the advantages of protection of anaerobes and more volume for sludge storage. The microbial metabolic activity in an anaerobic pond is similar to that of other anaerobic environments. Complex organic matter is broken down to short-chain fatty acids and alcohols by fermentation, and these materials are converted to methane gas by the methanogenic bacteria. Methane production is markedly influenced by temperature. Because temperature cannot be controlled, seasonal variations can be expected.

The main problem arising from anaerobic ponds is odor, which may be controlled by reducing the organic loading so that an aerobic layer becomes established at the pond surface. An alternative is to recirculate pond effluent back to the anaerobic pond. Pumping the recycled wastewater on the surface will aid in establishing an aerobic zone.

*R*EFERENCES

Atlas, R.M., and Bartha, R. (1987) *Microbial Ecology: Fundamentals and Applications.* 2nd Ed., Benjamin/Cummings, Menlo Park, Calif.

Benefield, L.D., and Randall, C.W. (1980) *Biological Process Design for Wastewater Treatment.* Prentice Hall, Inc., Englewood Cliffs, N.J.

Bowker, P.G., and Smith, J.M. (1985) Odor and Corrosion Control in Sanitary Sewerage Systems and Treatment Plants. EPA/625/1-85/018, U.S. EPA Technol. Transfer, Cent. Environ. Res. Inf., Cincinnati, Ohio.

Brock, T.D., and Madigan, M.T. (1991) *Biology of Microorganisms.* 6th Ed., Prentice Hall, Englewood Cliffs, N.J.

Gerardi, M.H. (1986) Control of Actinomycetic Foam and Scum Production. *Public Works.*

Hess, T.F., and Silverstein, J. (1990) Biodegradation of 2,4-Dinitrophenol in Sequencing Batch Reactors. Paper presented at 63rd Annu. Conf. Water Pollut. Control Fed., Washington, D.C.

Lemmer, H. (1986) The Cology of Scum Causing Actinomycetes in Sewage Treatment Plants. *Water Res.*, **20**, 531.

Metcalf & Eddy (1979) *Wastewater Engineering: Treatment, Disposal, Reuse.* 2nd Ed., McGraw-Hill, Inc., New York, N.Y.

Pike, E.B., and Curds, C.R. (1971) The Microbial Ecology of the Activated Sludge Process. In *Microbial Aspects of Pollution*, G. Sykes and F.A. Skinner (Eds.), Academic Press, New York, N.Y.

Sezgin, M., and Karr, P.R. (1986) Control of Actinomycete Scum on Aeration Basins and Clarifiers. *J. Water Pollut. Control Fed.*, **58**, 972.

Sims, A.F.E. (1980) Odor Control with Hydrogen Peroxide. *Prog. Water Technol.*, **12**(5), 609.

Speece, R.E., and Orosco, R. (1970) Design of U-Tube Aeration Systems. *J. Sanit. Eng. Div., Proc. Am. Soc. Civ. Eng.*, **96** (SA3), 715.

Stanier, R.Y., *et al.* (1986) *The Microbial World.* 5th Ed., Prentice Hall, Inc., Englewood Cliffs, N.J.

Stanier, R.Y., *et al.* (1966) The Aerobic Pseudomonads: a Taxonomic Study. *J. Gen. Microbiol.*, **43**(2), 159.

Strom, P.F., and Jenkins, D. (1984) Identification and Significance of Filamentous Microorganisms in Activated Sludge. *J. Water Pollut. Control Fed.*, **56**, 52.

U.S. EPA (1987) *The Causes and Control of Activated Sludge Bulking and Foaming*. EPA/625/8-87/012, U.S. EPA Technol. Transfer, Cent. Environ. Res. Inf., Cincinnati, Ohio.

U.S. EPA (1975) *Process Design Manual for Nitrogen Control*. Technol. Transfer, U.S. EPA Natl. Environ. Res. Center, Cincinnati, Ohio.

Water Pollution Control Federation (1990) *Wastewater Biology: The Microlife*. Special Publication, Alexandria, Va.

Chapter 5
Volatile Acids and Methanogenic Bacteria

Any naturally occurring material is biodegradable. Organic materials are present in waste products derived from home or industry, and may be simple or complex, depending on their chemical makeup. Most are lipids, carbohydrates, or proteins.

Lipids include fats, oils, and waxes. The breakdown products of fats and oils (*triglycerides*) are three fatty acid molecules (simple lipids) and one

---

molecule of glycerol, a sugar alcohol. The fatty acids are carbon chains and may vary in length. *Carbohydrates* are sugars and their derivatives. The simplest carbohydrates are the *monosaccharides* (single sugars), which include glucose, the energy source for living organisms. Simple sugars are linked together to form complex carbohydrates (*polysaccharides*) such as cellulose, a component of plant walls. The building blocks of proteins are amino acids, which may be single or joined by peptide bonds to form short or long chains (*polypeptides*).

The larger, more complex of these organic materials require a long period of time before their original mass is reduced by bacteria, while the smaller, simpler ones (short chain fatty acids, monosaccharides, and amino acids) can be decomposed in minutes or hours. When these waste products reach the treatment plant, they can include living or dead bacteria, protozoan, metazoan, or even dead human cells shed during bathing. Some of the organic compounds are *anthropogenic* (made by human beings) and they too can be decomposed to their elemental constituents as long as they are biodegradable.

All organic substances contain carbon, hydrogen, and oxygen and may include lesser amounts of nitrogen, phosphorus, or sulfur. The bulk of the products of complete decomposition of these substances can be carbon dioxide and water, ammonia, phosphates, sulfates, sulfides, or methane, depending on the original composition of the waste materials and whether they are treated aerobically or *anaerobically* (without oxygen). During biological decomposition, more than half of the organic compounds may be transformed to energy for the decomposers. The decomposers use the energy to synthesize more biomass.

An aerobic decomposition process, such as an activated sludge process, uses bacteria and other microbes. The oxygen that the operators supply is used by microbes in their respiration. They combine it with the hydrogen (H^+) released from the organic substrates to form water.

The focus of this chapter is on the anaerobic wastewater treatment process (*anaerobiosis*), with particular emphasis on fatty acid formation and methanogenesis. *Methanogenesis* is the conversion of low molecular weight fatty acids (volatile acids such as formic, acetic, propionic, and butyric acid) or alcohols, carbon dioxide, and hydrogen to methane gas (CH_4).

*V*OLATILE ACIDS FROM LIPIDS

All naturally derived lipids are biodegradable. Fats are not only of energy value to the anaerobes, but a number of fat-soluble vitamins (D, E, and K) that they require are often associated with fats. Some of the fats contain *unsaturated* fatty acids (some carbon atoms in the fatty acid chain are joined by double bonds) needed by the organisms. Lipids also include substances chemically related to fats, such as lecithin (a component of cell membranes) and cholesterols. Most lipids are nearly insoluble or slightly soluble in water but are completely soluble in alcohol-ether mixtures and related solvents (Holum, 1990). Carbohydrates and proteins are nearly insoluble in these solvents.

Lipids also may be neutral fats that are esters of fatty acids and glycerol. They also may be of several other types. Waxes are solid lipids consisting of three fatty acids and alcohols other than glycerol. Phospholipids typically contain fatty acids, glycerol, and a compound containing phosphorus, nitrogen, or both. Derived lipids are substances that originally were parts of fats and include fatty acids, straight-chain carboxylic acids, alcohols, and some vitamins.

Some of the more common fatty acids are presented in Table 5.1.

Table 5.1   Common fatty acids.

Fatty acid	Formula	Occurrence
Acetic	$CH_3 COOH$	Vinegar
Butyric	$C_3H_7 COOH$	Butter
Caproic	$C_5H_{11} COOH$	Butter
Caprylic	$C_7H_{15} COOH$	Butter
Capric	$C_9H_{19} COOH$	Coconut oil, butter
Lauric	$C_{11}H_{23} COOH$	Spermaceti, coconut oil
Myristic	$C_{13}H_{27} COOH$	Nutmeg butter, coconut oil
Palmitic	$C_{15}H_{31} COOH$	Animal and vegetable fats
Stearic	$C_{17}H_{35} COOH$	Animal and vegetable fats
Arachidic	$C_{19}H_{39} COOH$	Peanut oil

The fatty acids in Table 5.1 can form from the *hydrolysis* (a change in chemical composition produced by combination with water) of fats even in the sewer main and can react with metallic ions, such as sodium, potassium, magnesium, and calcium, to form salts or soaps (Harrow and Mazur, 1958). Metallic ions react with the acid group of fatty acids, rather than with the hydrocarbon chain. Sodium and potassium soaps are soluble in water, while magnesium and calcium soaps are insoluble. These insoluble soaps may compose the bulk of what is referred to as "scum" in the digester. So-called "hard" waters are those containing calcium and magnesium.

Unlike fats, fatty acids are readily broken down by a variety of hydrolyzing anaerobes through a complex biochemical pathway known as beta (β) oxidation, in which fatty acids are oxidized by successive removal of two-carbon fragments. These fragments are in the form of coenzymes. The removal of one particular coenzyme, called coenzyme A, results in the formation of fatty acids such as butyric, propionic, and acetic acid. These low molecular weight fatty acids are called volatile acids because they can *vaporize* (evaporate) at atmospheric pressure. Of these volatile acids, the most important precursor to methane is acetic acid, or acetate.

VOLATILE ACIDS FROM CARBOHYDRATES OR POLYSACCHARIDES

All sugars contain carbon, hydrogen, and oxygen and some may contain nitrogen and phosphorus. The basic chemical formula for sugars can be expressed as $(CH_2O)_x$. Monosaccharides (glucose, galactose, frutose) can enter one of a variety of biochemical pathways in the microbes. As a result of their simple structure and solubility in water, they can be transported across the microbial cell membrane with ease. *Disaccharides* (lactose, sucrose, maltose) are also water soluble but must be enzymatically hydrolyzed to monosaccharides before they can be transported across the cell membrane. Once inside the cell pathway, the particular metabolic pathway the sugar follows depends not only on the type of organism but also on whether that organism is in an aerobic or anaerobic environment.

Polysaccharides, the most complex sugars, can have a high molecular weight and are insoluble in water. Thus, their complete degradation to monosaccharides may require several enzymatic steps. For example, the common polysaccharides starch and cellulose are similar in chemical structure. These complex carbohydrates are first degraded by extracellular hydrolytic enzymes (*exoenzymes*) that are secreted by particular organisms. The hydrolysis of polysaccharides to disaccharides involves a unique enzyme for each specific reaction. For example, the enzyme amylase hydrolyzes starch to maltose, while the enzyme cellulase hydrolyzes cellulose to cellobiose. These disaccharides degrade to form many molecules of monosaccharides.

Monosaccharides can enter the anaerobic pathway known as glycolysis, which produces a versatile product called pyruvate or pyruvic acid, often known as the "hub" of biochemical reactions. It can enter almost any biological pathway, be it *anabolic* (synthetic) to make various cellular components, or *catabolic* (destructive) to form products such as formic and acetic acid (Frobisher *et al.*, 1974, and Stanier *et al.*, 1986).

The use of sugars by bacterial metabolic processes under anaerobic conditions is known as *fermentation*. Products of fermentation are alcohols (such as ethanol and butanol) that may be transformed to their corresponding fatty acids or volatile acids.

VOLATILE ACIDS FROM PROTEINS

Proteins are complex organic materials composed of simple building blocks known as *amino acids*. There are at least 26 amino acids in nature, although only 20 are common to most proteins. Some amino acids are straight chain (*aliphatic*) while others are *aromatic* (ring). Individually, they each contain an amino group ($-NH_2$), a carboxyl group (COOH), and a side group (designated "R"). Amino acids are joined by peptide bonds to form *polymers* (chains) of various lengths. These bonds are formed by removing an OH from a carboxyl group of one amino acid, and an H from the amino group of an adjacent amino acid. The bonding of two amino acids together form a *dipeptide*, with the OH^- and H^+ combining to form water. Therefore, when anaerobes degrade dipeptides or other proteins, water is added through hydrolysis, with OH^- going to one amino acid and H^+ to the other.

Many of the anaerobic bacteria in the digester are able to hydrolyze the proteins to either peptides of various chain lengths or to individual amino acids. The hydrolysis of proteins, similar to that of complex carbohydrates and lipids, requires the secretion of exoenzymes (proteases or peptidases) from these organisms. Free amino acids may undergo further anaerobic degradation (fermentation, decarboxylation, or deamination) and ultimately form volatile acids.

Volatile acids, therefore, can be derived from lipids, carbohydrates, and proteins. In living cells, biochemical pathways involving all three classes of food occur simultaneously and are highly coordinated, reducing energy loss from these substrates.

FACULTATIVE ANAEROBES, ANAEROBES, AND METHANOGENIC BACTERIA

All anaerobes are able to carry out life processes in the absence of dissolved oxygen. Thus, even in anaerobic conditions, such as in stored raw wastewater or activated sludge, simple and complex organic materials undergo various changes. Their chemical composition is altered based on the kind of bacteria that are metabolizing them.

The change in the microbial population in time in a given environment is called microbial succession and is an orderly process. For example, at the onset of succession, there may be an increase of versatile anaerobes that use a variety of complex organic compounds as food (faculative anaerobics). After the hydrolysis of complex organic substances by these organisms, most of the products formed are converted to various organic acids by anaerobic acid formers (*acetogenesis*). During this phase of anaerobiosis, low molecular weight fatty acids such as *formic* (one carbon), *acetic* (two carbon), *propionic* (three carbon), and *butyric* (four carbon) acids are formed.

Sulfate reduction (microbial reduction of sulfate to hydrogen sulfide, a malodorous and dangerous gas) precedes methanogenesis or biogas (CH_4) formation, with both being accomplished by a group of bacteria known as

---

methanogenic bacteria (Crowther and Harkness, 1983, and Stronach *et al.*, 1986). These bacteria are called *obligate* anaerobes because they can survive only in environments that lack oxygen. The anaerobic condition of an aquatic environment is expressed in terms of the oxidation-reduction potential, which is measured in millivolts (mV).

FACULTATIVE ANAEROBIC BACTERIA. Unlike obligate anaerobes, faculative anaerobes can carry out metabolic processes either in the presence of dissolved oxygen or the nearly total absence of it. They are, therefore, considered physiologically versatile organisms.

Many of the facultative anaerobic bacteria carry out mixed-acid fermentation. For example, the species of the genus *Enterobacter* can produce acids, aldehydes, alcohols, carbon dioxide, and hydrogen from a simple monosaccharide, glucose. Similarly, strains of *Escherichia coli* do nearly the same in addition to producing the malodorous intermediate known as indol or skatol, whose telltale odor is equated with stale, raw wastewater. Some of the facultative anaerobes are both hydrolyzers of complex organic substrates and acid formers.

ANAEROBIC BACTERIA. Unlike facultative anaerobic bacteria, anaerobic bacteria are unable to carry out their metabolic processes in an environment with dissolved oxygen. Most facultative anaerobes do well in aquatic environments having an oxidation-reduction potential between -200 and +200 mV without ill effects. Anaerobes, including methanogenic bacteria, however, can do better in an oxidation-reduction potential in the range -200 to -400 mV.

Anaerobes may be divided into two groups: those that cannot carry out metabolism if there is even a trace of dissolved oxygen but will at least survive the condition (oxygen-tolerant species) and those intolerant of any amount of dissolved oxygen in their environment. Such oxygen-intolerant anaerobes are killed in oxygenated environments and include all orders of methanogenic bacteria (Methanobacteriales, Methanococcales, and Methanomicrobiales). The terms obligate anaerobes and *strict*, or stringent, anaerobes are used to denote the intolerance of these organisms to dissolved oxygen (Brock, 1979).

Some anaerobes are strong acid producers (such as the anaerobic *Streptococcus*) and others reduce sulfate to hydrogen sulfide (species of *Desulfovibrio* and *Desulfomarculum*). The bulk of strict anaerobes are scavengers, inhabiting lakes and river bottoms, human intestinal tracts, or any other place where anaerobic conditions exist.

In stored wastewater or in stored activated sludge, facultative anaerobes and anaerobes (excluding methanogenic bacteria) carry out hydrolysis of complex organic substrates (carbohydrates, proteins, and lipids). Fats (triglycerides) are converted to fatty acids and glycerol, proteins are hydrolyzed to short peptide amino acids, and complex polysaccharides are converted to mono- or disaccharides. All of these metabolic intermediates are further converted to volatile acids, particularly acetic acid, and small molecules of alcohols, aldehydes, ketones, ammonia, carbon dioxide, hydrogen, and water (Mosey, 1983). Thus, hydrolyzers and acid formers prepare the environment for methanogenic bacteria both in terms of oxidation-reduction potential (-200 to -400 mV) and the production of precursor substrates for them to metabolize.

METHANOGENIC BACTERIA. Methanogenic bacteria are some of the most ancient forms of bacteria and are different from all other *eubacteria* (true bacteria). Some of the distinct characteristics of methanogenic bacteria include their unique metabolic pathways, cell wall composition, and coenzymes.

Methanogenic bacteria play an important role in nature and their environmental effect is substantial. They are able to convert such products of fermentation as formate and acetate to gaseous products (carbon dioxide, hydrogen, and methane) that can diffuse to the aerobic environment (atmosphere). This prevents the massive accumulation of organic materials that are considered *biorecalcitrant*, or that can only slowly be metabolized (Barker, 1956; Taylor, 1982; and Zeikus, 1977).

Methanogenic bacteria occupy a special niche in the microbial world because they alone produce a hydrocarbon (CH_4) as the major catabolic product. These bacteria are morphologically diverse, their shapes including long and short rods, filaments, lancet-shaped cocci, irregular clusters of small cocci, *sarcina* (cuboid arrangements of spherical cells), spirals, and curved rods. Some are *motile* (capable of movement), others nonmotile, and they may stain either gram-positive or gram-negative. The cell wall composition of methanogenic bacteria is such that it can be nonrigid. This characteristic was the first to distinguish this group from all other eubacteria.

According to *Bergey's Manual of Systematic Bacteriology* (1989), there are three orders of methanogenic bacteria divided into several families that include various genera consisting of more than 30 species.

There are five substrates that some species of methanogens convert to methane: acetate (CH_3-COO^-), formate ($HCOO^-$), methanol (CH_3OH), carbon dioxide (CO_2), and methylamine (CH_3NH_2). The use of acetate in methane formation involves a cleavage of the molecule, with the formation of methane from the -CH_3 group and of carbon dioxide from the carboxyl (COO^-) group. However, carbon dioxide produced from this cleavage can also be used as a substrate for methane formation. This reaction occurs by a reductive process.

O*RIGIN OF ANAEROBES IN ANAEROBIC DIGESTERS*

Anaerobes are found in both terrestrial and aquatic environments wherever anaerobic conditions exist. Many are found in the running water of rivers, but they are more prevalent in stagnant areas such as the bottom sediment of slow-flowing rivers, ponds, lakes, and oceans. Anaerobes are also common in soils and in the intestinal tract, thus feces, of humans and other animals. The intestine is an ideal place for anaerobes to thrive, with the numbers of certain anaerobes (such as species of *Bacteroides*) reported to be as high as 10^{10} cells/g of dry feces (Crowther and Harkness, 1983).

Some anaerobes are *saccharolytic* (sugar breaking) and are able to hydrolyze disaccharides and some trisaccharides and even larger polysaccharides to simple monosaccharides. These bacteria can number as high as 10^7 to 10^8 cells/mL in the digester and can include species in the genera *Clostridium, Acetoribrio, Staphylococcus, Bacteroides, Fusobacterium*, and *Peptococcus*. Similarly, there are *cellulolytic* (cellulose-breaking) anaerobes that hydrolyze cellulose and hemicellulose to monosaccharides. The number of these organisms range from 10^4 to 10^5 cells/mL in the digester liquor (Hobson and Shaw, 1974).

Anaerobic *proteolytic* (protein-breaking) organisms are responsible for hydrolyzing complex amino acid polymers, proteins, or polypeptides to simple amino acids. The number of bacteria performing anaerobic proteolysis, or putrefaction, can range from 10^5 to 10^6 cells/mL in the digester liquor (Kotze *et al.*, 1968, and Toerien *et al.*, 1970). These organisms include species in the genera *Clostridium, Proteins, Peptococcus, Bacteroides, Bacillus*, and *Vibro*.

Anaerobic *lipolytic* (lipid-breaking) organisms hydrolyze saturated or unsaturated fats, oils, and waxes to various fatty acids and volatile fatty acids. Their numbers are in the range of 10^4 to 10^5 cells/mL in the digester liquor. Examples in this group include species of *Clostridium, Sarcia*, and *Staphylococcus*.

Thus, while their detailed biochemical activities may differ, some species of the same genera are able to hydrolyze more than one group of substrates. Methanogenic bacteria are also found in the digester, their numbers typically in the range of 10^4 to 10^8 cells/mL (Siebert *et al.*, 1968, and Smith, 1966).

*A*NAEROBIC DIGESTION OF ORGANIC MATERIALS TO METHANE

The sanitary or environmental microbiologist frequently divides the anaerobic digestion of complex organic molecules to methane into stages, the number of which may vary. Some simply divide the process into two stages: hydrolysis and volatile acid formation and methane formation. Others divide it into three stages, with hydrolysis of organic polymers as a separate first step.

Still other microbiologists divide the process into six stages. For example, stage one is the hydrolysis of proteins, lipids, and carbohydrates; stage two is fermentation or acid formation (*acetogenesis*) from products formed during stage one; stage three is the anaerobic hydrolysis or oxidation of fatty acids (β-oxidation) and alcohols; stage four is the anaerobic oxidation of fatty acids and some volatile acids (propionic and butyric); stage five is the conversion of acetate-to methane; and stage six is the coupling of hydrogen and carbon dioxide to methane (Stronach *et al.*, 1986).

These steps are convenient for outlining individual biochemical activities that can be carried out by a variety of anaerobic bacteria such as hydrolyzers, acetogenic bacteria, and methanogens. However, the reaction may not involve many different anaerobics because a single species is able to carry out the first four stages (covering the conversion of organic materials to acetate). Stages five and six are carried out by methanogenic bacteria.

*F*ACTORS INFLUENCING ANAEROBIOSIS AND METHANOGENESIS

Various environmental factors may affect the efficiency of anaerobic digestion and methanogenesis. Among these are

- Composition of nutrients;
- Dissolved oxygen content of the anaerobic digester;
- Temperature of the digester;
- pH (hydrogen ion concentration) of the digester;
- Concentration of volatile solids; and
- Concentration of volatile acids.

COMPOSITION OF NUTRIENTS. Anaerobic processes can be used to treat high-strength raw wastewater containing low or high amounts of suspended solids. These processes can also be used to treat primary and secondary sludge derived from conventional aerobic treatment processes, such as those from municipal or industrial facilities. However, with few exceptions, wastewater and sludge are heterogeneous mixtures of complex organic materials differing widely in their chemical and physical composition. Thus, it is best to report "average composition" as may be found in various references and guides (Fisher and Swanwick, 1971; Malina, 1971; Swanwick and O'Gorman, 1973; and Swanwick *et al.*, 1969). For example, it is reported that sludge solid content of facilities in the U.S. may vary from 4 to 6% on a dry weight basis.

However, a more important parameter than dry weight in stabilizing waste materials by anaerobic digestion is total organic matter. This is a measure of materials including complex carbohydrates, proteins, and lipids. All of these substrates can be converted to volatile acids, thus to methane.

Faculative and strict anaerobes involved in anaerobic digestion obtain both energy and materials needed to synthesize cellular components from the sludge. In addition, the substrate (sludge) should contain vitamins and other necessary growth factors. Many of these are exogenously provided by operators to the sludge while others can be synthesized by the bacteria themselves in the digester. For example, starch may be enzymatically hydrolyzed to a disaccharide (maltose), and subsequently to monosaccharide (two molecules of glucose) units.

This degradation is necessary because most of the mono- or disaccharides, which may have been present in the raw wastewater, have already been consumed during the aerobic treatment process or possibly during the primary sedimentation. The sugars derived from hydrolysis are converted to volatile acids, carbon dioxide, and hydrogen. Complex proteins and peptides are similarly digested through *proteolysis* (hydrolysis) to short peptides and amino acids. These products are further digested (decarboxylated and deaminated) to various organic acids or alcohols and, ultimately, to volatile acids.

During hydrolysis, amino groups, amino acids, and all other nitrogen-containing organic compounds such as deoxyribonucleic acid (DNA), ribonucleic acid (RNA), and enzymes are *deaminated* (their $-NH_2$ groups are enzymatically removed). This results in an increase in the total nitrogen content of the sludge, with ammonia (NH_3)—a byproduct of deamination—being converted to ionized ammonium (NH_4^+) (Cooney, 1981). At this stage, fats and grease can be converted to fatty acids of various chain lengths. The long-chain fatty acids are further hydrolyzed to smaller fatty acids (volatile acids) (Hobson, 1982, and Van Assche, 1982). Those nutrients that should be exogenously provided to the sludge include carbon, nitrogen, phosphorus, sulfur, metal ions, and any other elements needed by the microbes to carry out their anaerobic life processes.

The composition of nutrients in the sludge also influences the composition of biogas. As a rule, CH_4 represents the greatest portion of offgas (65 to 90%) in the digester. Although the percentage of carbon dioxide may be greater, it is found mostly in the dissolved state in the digester liquor. Because the final stage in anaerobic digestion is methane production, the performance of a digester can be expressed in terms of the amount of methane released per unit of raw sludge. A complete carbohydrate breakdown can yield 50 parts methane and 50 parts carbon dioxide. The contribution of proteins to gas production is not as well documented, although lipid digestion gives a proportionately greater amount of methane than do either proteins or carbohydrates. To achieve optimum digester performance, the feed stream should contain all foods (including vitamins and mineral elements) that are necessary for the completion of hydrolysis, acetogenesis, and methanogenesis (Lettinga *et al.*, 1981).

DISSOLVED OXYGEN CONTENT OF THE ANAEROBIC DIGESTER.
Any amount of dissolved oxygen in the anaerobic digester can discourage hydrolysis, acetogenesis, and methanogenesis. As a rule, wastewater sludge that has undergone sedimentation and thickening is nearly anaerobic, meaning it has a low oxidation-reduction potential (-100 to -300 mV).

TEMPERATURE OF THE DIGESTER. Almost all biological activities are influenced by temperature. The temperature of the digester can enhance or inhibit specific growth rate of microbes, decay rate of materials, gas production, substrate use, and many other biological activities. One reason for this is that these activities involve many enzyme-mediated reactions, and enzymes are temperature sensitive. For example, with every 10°C rise in temperature (within the minimum and maximum growth temperature range of the microbes), the activity of an enzyme doubles. Therefore, it can take twice as long for a substrate to be digested at 25°C (75°F) as it would at 35°C (95°F), providing all other factors operating in the digester are the same (Switzenbaum and Jewell, 1980; and Zinder *et al.*, 1984).

As a result of the influence of temperature on biological activity and decomposition rate, the temperature of the digester should change with the age of the sludge.

Sludge age should decrease with increasing temperature. For example, at a temperature of 18°C (65°F), the sludge age should be 28 days, whereas at a temperature of 30°C (85°F), the recommended sludge age is reduced to 14 days. One reason for this temperature variation is that different groups of anaerobes are active at different times over the course of digestion, and each of these groups has its own temperature optimum for maxium growth and metabolism. Anaerobes can be classified according to the temperature range over which they are most active. Three general groups are recognized: mesophiles, psychrophiles, and thermophiles.

Mesophilic Anaerobic Digestion. Most of the anaerobic digesters used in the U.S. operate at a mesophilic range of between approximately 20 and 45°C (70 and 115°F). This is practical because there are more anaerobic mesophiles in nature than there are psychrophiles and thermophiles. Also, it is easier and less costly to maintain mesophilic temperatures in digesters than it is to maintain psychrophilic or thermophilic temperatures.

The sludge retention time (SRT) in a mesophilic digester also is shorter than it is in a psychrophilic digester (4 to 6 weeks compared to 12 weeks), thus anaerobiosis occurs faster.

Psychrophilic Anaerobic Digestion. Sludge digestion at this temperature—5 to 25°C (40 to 75°F)—typically is confined to a small-scale operation or to facilities such as septic tanks, imhoff tanks, and sludge lagoons where the digester is unheated and the temperature of the material in it is nearly equal to that of the outdoor environment. If the digester is unheated, the rate of digestion varies from season to season with the temperature of the digester contents. The anaerobes in such a system also experience seasonal change, or turnover, with changing temperatures. Because temperatures remain relatively low, the SRT of these digesters can range from 100 to 400 days or longer (Schraa and Jewell, 1984).

Thermophilic Anaerobic Digestion. Because the rate of anaerobic digestion and methanogenesis is proportional to digester temperature, the degradation rate of thermophilic digesters, which operate between approximately 50 and 70°C (120 and 160°F), is considerably faster than in a mesophilic system. Thus, the loading of thermophilic digesters may be high, but the temperature control is more difficult and expensive in these digesters than in either the psychrophilic or mesophilic digesters.

In the mesophilic, psychrophilic, and thermophilic digestion of sludge, it is important to maintain a uniform temperature throughout the vessel. Any localized pocket temperature or variation among areas of the digester may inhibit or inactivate certain anaerobic bacteria, including methanogens acclimated to a narrow temperature range (Mosey, 1983). For example, if a mesophilic digester containing *Methanobacterium bryantii* (37 to 39°C [99 to 102°F]), *Methanobacterium alcaliphilum* (37°C [99°F]), and *Methanobacterium uliginosum* (37 to 40°C [99 to 104°F]) as the principal CH_4 producers were operating at 35°C (95°F), a 10°C temperature increase can stop methane production within a period as short as 12 hours. A temperature change of even a few degrees tends to affect almost all biological activities, making it necessary to maintain adequate mixing of the digester contents to avoid localized temperature variations.

pH OF THE DIGESTER. Because enzyme activity is significantly influenced by pH, any change in pH carries important implications for biological systems. Most organisms have a minimum, maximum, and optimum pH for growth and reproduction, and microbiologists can classify groups of microbes according to their optimum pH ranges. *Neutrophyles* are those organisms preferring a pH between 6 and 8.5. *Acidophyles* prefer a pH between 3.5 and 5.5, and *Basophyles* prefer one between 9.5 and 13. These values are approximate because, for some organisms, the pH range is narrow while for others it is broad.

Because of its influence on enzyme activity, the control of pH is important in establishing optimum growth and reproductive conditions for organisms. These activities are related to various metabolic processes, including hydrolysis, volatile acid formation, and methanogenesis. Most organisms in the anaerobic digester do well between a pH of 6.5 and 7.5. This narrow but optimum pH range is largely controlled by the buffering capacity of the digester liquor. *Buffering capacity* represents the ability of a liquid to resist a change in pH. Two natural buffering agents operating in the digester are ammonium (NH_4^+) and bicarbonate (HCO_3^-). Ammonium ions are largely derived from the deamination of amino acids and other nitrogen or amine-containing cellular materials, such as protein, DNA, RNA, adenosine triphosphate, and enzymes. Bicarbonate ions are derived from carbon dioxide produced during hydrolysis, acid formation, and methanogenesis.

The bicarbonate ion concentration (sometimes referred to as bicarbonate alkalinity) is important in pH control in the anaerobic digester because it can suppress the effects of high hydrogen ion concentrations (low pHs). If the pH of the digester media deviates too far from neutral in either direction, anaerobic digestion may be inhibited. Thus, for maximum biological activity, it is best to try to maintain a pH of 7.0 (neutral) in the digester.

CONCENTRATION OF VOLATILE SOLIDS. The fraction of volatile solids in the digester is an important parameter and can be used in the calculation of loadings. The higher the volatile solid concentration of the feed, the higher the loading. Volatile solids are food sources for hydrolytic and acid-forming anaerobes in the digester. These solids, along with carbon dioxide and water, are also the sources of volatile acids (formic, acetic, propionic, and butyric acids) for methanogenic bacteria. Therefore, the fraction of volatile solids in the digester is directly related to the amount of volatile acids formed, and has a profound influence on the hydrogen ion concentration (pH) of the digester.

CONCENTRATION OF VOLATILE ACIDS. Wastewater from pretreated food-processing industries—especially those using unsaturated fats or oils—can contain large amounts of fats, grease, and lower molecular weight fatty acids (acetic, propionic, and butyric acids). Such pretreatment wastewater can contain not only volatile acids, but also high-molecular weight fatty acids derived from common fats (triglycerides). Large amounts of fatty acid can neutralize the bicarbonate alkalinity (bicarbonate ion concentration) built up in the digester (Ianotti and Fischer, 1984).

When digested, fatty acids can be converted to volatile acids, an excessive accumulation of which can cause a severe drop in the pH of the digester liquor. Such a decrease in pH would inhibit both hydrolysis and methanogenesis (Marty and Germe, 1986). Acetogenic bacteria are more tolerant of a low pH, but also are affected if the pH of the digester drops to 4.5 or lower. Thus, a sudden or dramatic increase in loading should be avoided because of the volatile acid formation that will follow. Any increases should be made slowly, with careful consideration of the buffering capacity of the digester.

*C*OMMON SUBSTANCES TOXIC AND INACTIVATING TO DIGESTER MICROBES

Some substances in moderate to excessive amounts can cause problems in anaerobic digestion, including heavy metal ions, sulfide, dissolved ammonia gas, un-ionized volatile acids, and cyanide.

HEAVY METAL IONS. Certain metals can become toxic if present in large concentrations in the anaerobic digester. Although some metals such as zinc, copper, nickel, cobalt, boron, molybdenum, selenium, iron, magnesium, and manganese are considered essential in carrying out life processes, they can inactivate many anaerobes, including methanogens, if their concentrations exceed the optima. Some of the more influential of these heavy metal ions are mercury, cadmium, chromium, lead, copper, and nickel, and their toxicity to bacteria is well documented (Duarte and Anderson, 1983). These soluble heavy metals (ions) serve as inhibitors of microbial enzymes. In the digester, toxic levels of these metal ions vary considerably with the chemical composition of the substrates to be digested.

The toxic effects of heavy metal ions may become masked in the digester. For example, many of the metal ions can be precipitated by sulfide (S^{2-}) and carbonate (CO_3) ions reacting with the metal ions to form insoluble salts of sulfides or carbonates.

The toxic effects of heavy metal ions can also be masked by organic acids. These acids, resulting from substrate digestion, can render the ions inert through a process called *chelation*. Both precipitation and chelation are often controlled by the pH of the digester liquor. Any pH greater than 7.5 would promote better precipitation of the salts of carbonates or sulfides. In this precipitated form, the bacteria can better tolerate the metal ions, even if the concentration of ions is great.

Most of the sulfide present in an anaerobic digester is derived from sulfates (SO_4) present in the feed water. These sulfates become reduced to hydrogen sulfide in an anaerobic environment. Other sources of sulfide are the proteins in the sludge. Some amino acids (cystine, cysteine, and methionine) in these proteins contain sulfur, which is released on digestion in the form of a thiol group (-SH).

A third means of reducing the toxicity of heavy metal ions is through their adherence to the surface of influent-suspended solids (*adsorption*). Thus, the higher the concentration of these solids, the more metal ions can be adsorbed, provided the adsorption sites are not occupied by metal ions before entering the digester.

Heavy metal ions become toxic to microbes by rendering their enzymes inactive. This is accomplished by the metal ions combining with the -SH group of the enzymes. When this occurs, digestion and methane formation become impossible (Parkin and Speece, 1983). The concentrations at which heavy metal ions become inhibitory depends on the chemical composition of waste materials fed to the digester.

In part, heavy metal toxicity depends on whether the metal salts in the waste are soluble or insoluble. If the salts are soluble, as are those of chloride and nitrate, they will *ionize* (dissolve) in aquatic systems. The ionized metal ions are then free to react with enzymes and other essential components of living cells, thus interfering with digestion and methane production.

SULFIDE INHIBITION AND TOXICITY. The formation of sulfide is an inevitable consequence of the anaerobic reduction of sulfate (SO_4^{2-}) and of the decomposition of sulfur-containing amino acids. Anaerobic reduction of sulfate or even the reduction of elemental sulfur to hydrogen sulfide is carried out in any environment that lacks oxygen, such as an anaerobic digester.

Hydrogen sulfide formation in sewer mains also is common, and many workers in the environmental sanitation field have lost their lives by inhaling this toxic gas. Sulfides formed from sulfate- or sulfur-containing amino acids may be soluble or insoluble in the digester liquor. Insoluble sulfides have little effect on the performance of the anaerobic digester. These compounds include lead sulfide (PbS) and iron sulfide (Fe_2S_3). In fact, iron has been added to digesters, where it combines with sulfide to alleviate its toxicity. The iron sulfide precipitates and imparts a black coloration to the treated sludge.

Many anaerobic bacteria can remove the thiol (-SH) group from sulfur-containing amino acids and use it for building their own cells. However, there are approximately seven genera of anaerobic bacteria that reduce sulfate or sulfur to dissolved sulfide (H_2S). The genus name of these sulfur-reducing organisms begins with the prefix "Desulf" (*Desulfuromonas, Desulfovibrio, Desulfomonas*). Similar to methanogenic bacteria, their habitat is anaerobic. Their cellular morphology also is similar to that of methanogenic bacteria in its diversity.

The mechanism by which the toxic, dissolved sulfide inhibits the metabolic processes of the anaerobes is not completely understood. However, its inhibitory or toxic effects can occur at concentrations as low as 200 mg/L in a mesophilic digester. Dissolved sulfide can react with any of the heavy metals, except chromium, in the digester and precipitate. Thus, it may function as a chelator of toxic heavy metals. Free sulfide (H_2S gas) can be removed from the digester by the vigorous production of other gases such as carbon dioxide, hydrogen, and methane.

DISSOLVED AMMONIA GAS. The toxic effect of ammonia may be confined to methanogens, but moderate amounts have some effect on hydrolytic and acid-forming organisms. The precise level at which ammonia is toxic remains uncertain. Some investigators have found that ammonia concentrations as great as 7 000 mg/L have little effect on methanogenesis, while others report inhibition at concentrations as low as 150 mg/L.

$$\underset{NH_3}{\underset{\text{Ammonia}}{}} \quad \rightleftarrows \quad \underset{NH_4^+}{\underset{\text{Ammonium ion}}{}} \quad + \quad \underset{OH^-}{\underset{\text{Hydroxide ion}}{}} \quad\quad (5.1)$$

As pH decreases from 7.0 to 5.5 (hydrogen ion concentration increases), the equilibrium of the reaction shifts to the right, with more ammonium ions being formed. The reverse is true as pH increases (hydrogen ion concentration decreases), with a shift to the left producing more ammonia.

Ammonia can be an important buffer in an anaerobic digester. The buffering capacity of ammonia allows the digester system to be self-regulating. For example, if the concentration of dissolved ammonia is great enough to cause methanogenic inhibition, an accumulation of volatile acids results. This acid accumulation causes a depression of the pH value, converting the dissolved ammonia to less toxic ammonium ions (NH_4^+). Thus, the inhibition of methanogenic bacteria is alleviated.

Operators may be able to detect problems caused by high concentrations of dissolved ammonia by noticing a decrease in methane production and an increase in volatile acid accumulation.

One of the more common causes of digester failure at elevated ammonia concentrations is inadequate *acclimation* (physiological adjustment) of methanogenic bacteria to such increases. However, many bacteria are capable of physiologically adjusting to gradual changes in environmental variables. Thus, anaerobes that are slowly acclimated may better tolerate greater ammonia concentrations.

A second possible reason for digester failure is shock loading. This phenomenon occurs when high ammonia concentrations are added to a digester operating at or over its design limit. Shock loading causes a rapid and excess production of volatile acids, resulting in a large drop in pH. Consequently, the buffers in the digester are unable to compensate for this sudden change in pH, causing digester failure. The precise concentration of ammonia or ammonium ions at which methanogens become inhibited remains uncertain. Variations occur among digesters with the nature of the waste to be treated, the environmental conditions surrounding a digester, temperature, loading rate, and buffering capacity. With a proper loading rate and acclimated anaerobes, however, toxicity can be reduced.

UN-IONIZED VOLATILE ACIDS. A high accumulation of low-molecular weight fatty acids such as butyric, propionic, and acetic acid (volatile acids) causes a drop in the pH of the digester contents. Such an accumulation of dissolved volatile acids can stem from a variety of causes, including failure of methanogens to convert these acids to methane, shock loading of the digester with volatile acids, and the infiltration of inhibitory substances such as heavy metals, chlorinated hydrocarbons, cyanides, formaldehydes, or chloroform.

Among the volatile acids, acetate appears to be the least toxic to most methanogenic bacteria in the digester, followed by butyric acid (McCarty and Brosseau, 1963). Concentrations of bacteria as great as 10 mg/L can be tolerated by digester microbes, including methanogens. High concentrations of propionic acid (1 to 5 mg/L), however, frequently have been cited as a cause of digester failure.

The problem of excessive accumulation of volatile acid ions can be solved by the addition of an alkaline (basic) substrate, such as sodium, hydroxide, or calcium carbonate (lime).

CYANIDE. Although the mechanism is unknown, cyanide and cyanide-containing compounds have been known to completely inhibit methanogenesis. For example, methanogens exposed to cyano-compounds for 24 hours at concentrations of 100 mg/L can stop production, though the performance of the system returned to normal in 5 days, indicating that the cyanide poisoning was reversible. Other studies indicated that exposure to cyano-compounds in concentrations less than 10 mg/L for approximately 1 hour have caused little inhibition of anaerobes in digesters. Thus, the presence of cyanide in digesters is not always completely inhibitory.

OTHER INHIBITORS. Some common household detergents can also affect the performance of organisms in the anaerobic digester (Mosey, 1983). For example, such compounds as lauryl sulfate have been found to burst the cell walls of both gram-positive and gram-negative bacteria. Thus, the operator can anticipate problems if the facility treats industrial waste containing large concentrations of detergents.

Also, there are several dozen chlorinated, anthropogenic organic compounds that can enter wastewater treatment facilities and become inhibitory or even toxic to digester organisms. These compounds can originate from a variety of sources, including *solvents* (dissolving agents), dry cleaning agents, pesticides, herbicides, insecticides, fungicides, and wood preservatives. They have been found to inhibit both methane formation and the growth of methanogenic bacteria, although their mode of action is not clear. To anticipate potential problems, operators should be aware that there will be increased amounts of chlorinated compounds and cyano-compounds introduced to digesters as the municipality accepts more and more industrial wastewater.

REFERENCES

Barker, H.A. (1956) *Bacterial Fermentations*. John Wiley and Sons, New York, N.Y.

Bergey's Manual of Systematic Bacteriology (1989). Volume 3, J.T. Staley *et al.* (Eds.), Williams and Wilkins, Los Angeles, Calif.

Brock, T.D. (1979) *Biology of Microorganisms*. Prentice Hall, Inc., Englewood Cliffs, N.J.

Cooney, C.L. (1981) Growth of Microorganisms. *Biotechnology, a Comprehensive Treatise in 8 Volumes*. In Volume 1, G. Reed and H.J. Rehm (Eds.), Yerlag Chemie, Weinheim-Deerfield Beach, Florida Basel.

Crowther, R.F., and Harkness, N. (1983) Anaerobic Bacteria. In *Ecological Aspects of Used-Water Treatment*. Volume 1, C.R. Curds and H.A. Hawkes (Eds.), Academic Press, London, U.K.

Duarte, A.C., and Anderson, G.K. (1983) Chem. Eng. Prog. Symp. Ser., **77**, 149.

Fisher, W.J., and Swanwick, J.D. (1971) *Water Pollut. Control*, **70**, 355.

Frobisher *et al.* (1974) *Fundamentals of Microbiology*. W.B. Saunders, Philadelphia, Pa.

Harrow and Mazur, (1958) *Textbook of Biochemistry*. W.B. Saunders, Philadelphia, Pa.

Hobson, P.N. (1982) Production of Biogas from Agricultural Wastes. In *Advances in Agricultural Microbiology*. N.S. Rao and Subba (Eds.), Butterworth Scientific, London, U.K.

Hobson, P.N., and Shaw, B.J. (1974) *Water Res.*, **8**, 507.

Holum, J.R. (1990) *Fundamentals of General, Organic and Biological Chemistry*. John Wiley & Sons, New York, N.Y.

Ianotti, E.L., and Fischer, J.R. (1984) Effects of Ammonia Volatile Acids, pH and Sodium on Growth of Bacteria Isolated from a Swine Manure Digester. In *Developments in Industrial Microbiology: Proc. 40th Gen. Meeting Soc. Ind. Microbiol.*, Sarasota, Fla.

Kotze, J.P., *et al.* (1968) *Water Res.*, **2**, 195.

Lettinga, G., *et al.* (1981) *Water Res.*, **15**, 171.

Malina, J.F., Jr. (1971) In *Manual of Wastewater Operations*. C.H. Billings and D.F. Smallhorst (Eds.), Tex. State Dep. of Health, Austin, Tex., 319.

Marty, B., and Germe, S.A. (1986) Microbiology of Anaerobic Digestion. In *Anaerobic Digestion of Sewage Sludge and Organic Agricultural Wastes*. A.M. Bruce *et al.* (Eds.), Elsevier Appl. Sci. Publishers, London, U.K.

McCarty, P.L., and Brosseau, M.N. (1963) Effects of High Concentration of Individual Volatile Fatty Acids on Anaerobic Treatment. *Proc. 18th Ind. Waste Conf., Purdue Univ.*, West Lafayette, Ind.

Mosey, F.E. (1983) Anaerobic Processes. In *Ecological Aspects of Used-Water Treatment*. Volume 2, C.R. Curds and H.A. Hawkes (Eds.), Academic Press, London, U.K.

Parkin, C.F., and Speece, R.E. (1983) *Water Sci. Technol.*, **15**, 261.

Schraa, G., and Jewell, W.J. (1984) High Rate Conversions of Soluble Organics with a Thermophilic Anaerobic Attached Film Expanded Bed. *J. Water Pollut. Control Fed.*, **56**, 226.

Siebert, M.L., *et al.* (1968) *Water Res.*, **2**, 545.

Smith, P.H. (1966) *Dev. Ind. Microbiol.*, **7**, 156.

Stanier, R.Y., *et al.* (1986) *The Microbial World*. Prentice Hall, Inc., Englewood Cliffs, N.J.

Stronach, S.M., et al. (1986) *Anaerobic Digestion Processes in Industrial Wastewater Treatment*. Springer-Verlag, Berlin, Germany.

Swanwick, J.D., *et al.* (1969) Name of article. *Water Pollut. Control*, **68**, 639.

Swanwick, J.D., and O'Gorman, J.V. (1973).

Switzenbaum, M.S., and Jewell, W.J. (1980) Anaerobic Attached-Film Expanded-Bed Reactor Treatment. *J. Water Pollut. Control Fed.*, **52**, 1953.

Taylor, G.T. (1982) *Prog. Ind. Microbiol.*, **16**, 231.

Toerien, D.F., *et al.* (1970) *Water Res.*, **4**, 129.

Van Assche, P.F. (1982) *Antonie van Leeuwenhoek*, **48**, 520.

Zeikus, T.G. (1977) *Bacteriol. Rev.*, **41** (2), 514.

Zinder, S.H., *et al.* (1984) *Appl. Environ. Microbiol.*, **47**, 808.

Chapter 6
Metabolic Processes

GENERAL PROCESSES

Metabolism is the sum of all transformations that occur in a cell such as that of a bacteria or protozoa. The purpose of these transformations is to create new cell mass so that cells can grow and divide and the microbial population can prosper. The general processes common to all cells that collectively comprise metabolism include energy conservation, biosynthesis, assimilation and ingestion, and cell maintenance.

The processes operate together to perform all necessary cellular functions. Figure 6.1 illustrates the scheme by which these processes interact to promote growth and sustain the viability of a microbial culture.

ENERGY CONSERVATION. The ability to transform and consume energy is a fundamental property of all living organisms. Cells require energy to perform mechanical work, such as locomotion, and electrical work, such as the movement of ions and molecules. Chemical energy also is required for and is

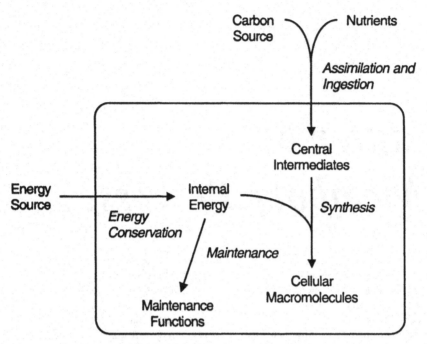

Figure 6.1 General overview of microbial metabolism.

consumed during reactions that synthesize cellular components, particularly macromolecules. Thus, energy consumption is essential to a cell's survival. Energy is derived from either chemicals called substrates, or from solar radiation. In wastewater treatment, the substrates are the wastewater constituents.

Organisms that use organic or inorganic molecules as their primary source of metabolic energy are called *chemotrophs*. Organisms that use light as their primary energy source are called *phototrophs*. For both types of organisms, it is essential that useful chemical energy be derived from the primary energy source through some transformation, or reaction. Invariably, some of the primary energy is converted to thermal energy, or heat, by such transformations and is lost to the environment. Ideally, however, a substantial portion of the energy from the primary source is *conserved* in such a way that it is available to the organism for cellular work, cell maintenance, synthesis of new molecules, and reproduction.

Energy Carriers. Figure 6.1 illustrates the importance of energy conservation in the overall metabolic scheme. The most common manifestation of this conservation of energy is through the manufacture of high-energy phosphate bonds, a form of chemical energy. Upon *cleavage* (breakage) of these bonds, a substantial amount of energy is made available to perform useful chemical, mechanical, or electrical work. The chemical compound that most often serves as a carrier of these high-energy bonds is adenosine triphosphate (ATP), the molecular structure of which is shown in Figure 6.2. The molecu-

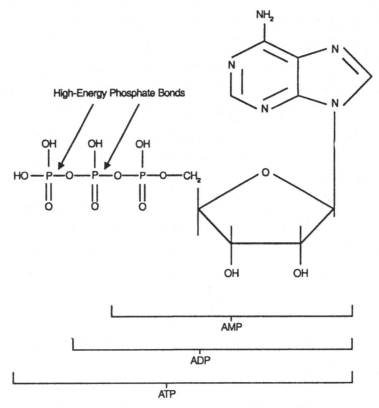

Figure 6.2   Adenosine, mono-, di-, and triphosphate molecules (Brock *et al.*, 1984).

lar product of bond cleavage, either adenosine diphosphate (ADP) if one bond is cleaved, or adenosine monophosphate (AMP) if two are cleaved, will serve as a base for the further production of ATP in energy-conserving reactions. Thus, metabolic energy is continuously produced and used through the breaking and formation of the phosphate bonds of ATP.

Anabolism is a constructive form of metabolism that uses energy (ATP) to synthesize more complex cellular components (macromolecules) from simpler compounds in the cell (central *intermediates*). The new cellular components are *phosphorylates*, meaning they gain a phosphate molecule (PO_4) from ATP in the process.

There are various mechanisms by which organisms conserve energy with the production of ATP. All mechanisms, however, operate by the common principle of oxidation-reduction, a class of chemical reactions characterized by the transfer of electrons from a donor molecule to an acceptor molecule. In these reactions, the donor molecule is oxidized, or gives up hydrogen atoms (electrons), while the acceptor molecule is reduced, or receives hydrogen atoms (electrons). This process is a fundamental way by which living organisms use the energy present in molecules for the performance of useful work.

Electron Carriers. Because the way in which organisms conserve energy is through oxidation-reduction reactions, the function of the primary energy source is to initiate the donation of electrons. Chemical substrates do so directly, by serving as electron donors. Sunlight does so indirectly, by causing light-receiving (*photoreceptor*) molecules such as chlorophyll-*a* to give up (donate) electrons. The donated electrons are originally received (accepted) by an electron-carrying organic molecule called a *coenzyme*. Coenzymes assist enzymes in speeding up chemical reactions.

The most common electron-carrying coenzyme is nicotinamide adenine dinucleotide (NAD), and its phosphorylated derivative, NADP. Nicotinamide adenine dinucleotide is a coenzyme for enzymes that speed up oxidation-reduction reactions. Upon the acceptance of a *hydride* ion (a hydrogen atom and two electrons), this molecule is reduced to NADH (NADPH). The reduced molecule then donates the electrons toward the production of ATP (in certain cases) or toward performing other chemical reductions required by the cell, such as assimilation, biosynthesis, or maintenance. Once the NAD molecule donates its hydride ion (electrons) toward any of these goals, it is again available to accept electrons. Thus, NAD is analogous to ATP in that it is a universal carrier of electrons (reducing power) in the same way that ATP is a universal carrier of phosphate bonds (metabolic energy).

BIOSYNTHESIS. The majority of the organic material in organisms consists of four classes of complex macromolecules: nucleic acids, proteins, polysaccharides (carbohydrates), and lipids (fats).

These large molecules are synthesized from a pool of simple organic compounds in the cell called central intermediates, which are products derived from the principal carbon source. The process of macromolecular formation from central intermediates is called biosynthesis, or anabolism, and can be broken down into two major steps: synthesis of precursor monomers and polymerization.

Precursor monomers (compounds composed of single molecules) serve as building blocks for the formation of macromolecules; that is, the macromolecules are chains (polymers) of precursor molecules formed by chemical bonding of the monomers. Both steps involved in biosynthesis are net consumers of metabolic energy and will use ATP formed during energy conservation. In the absence of sufficient amounts of ATP, biosynthesis will cease.

ASSIMILATION AND INGESTION. Organisms synthesize new cell mass from simple molecules through the use of metabolic energy made available by cleaving phosphate bonds in ATP. For this to occur, the constituents necessary for biosynthesis must be made available to the internal cellular environment. These constituents include a carbon source, nutrients, and the energy source, if this happens to be a chemical compound (as opposed to light). Growth nutrients and inorganic energy sources are always dissolved constituents. Most organic compounds used for carbon and energy sources are also in a dissolved state. The transport of *soluble* (dissolved) metabolic constituents from the external to internal environment is known as *assimilation*. Cells that obtain all of their metabolic constituents in this manner are known as *osmotrophs*.

In many environments, however, there exists a large amount of insoluble (undissolved) organic matter not available for direct assimilation by osmotrophs. This *particulate* (solid) organic matter may be solubilized through the action of exocellular enzymes or exocellular products of metabolism (such as acids), which can be released by organisms. The solubilized organic matter is then available for assimilation by osmotrophic cells. Certain organisms do have the ability to directly transport insoluble organic matter across their cell membranes, a process called *ingestion*. Cells capable of ingesting particulate organic matter are called *phagotrophs*.

Assimilation. Assimilation may be accomplished by either passive (energy-neutral) or active (energy-consuming) transport mechanisms. *Passive transport* across the cell membrane is caused by molecular *diffusion*, whereby a spatial difference in the concentration of a molecule or ion causes its net movement from regions of high to regions of low concentration. The process is *spontaneous* (does not require the input of work) and will continue until a uniform concentration between adjacent regions is attained. Thus, assimilation by passive transport can occur only when there exists, in the external environment, a higher concentration of the required components than is present inside the cell.

The cell membrane can act as a barrier to passive transport of selected components (be selectively *permeable*). Thus, passive transport is possible only when molecules able to diffuse through the cell membrane are present at excessive levels.

Most molecules of importance to cellular metabolism and behavior tend to occur intracellularly at levels higher than those in the adjacent, extracellular environment. Thus, most organisms must have the ability to assimilate compounds against a concentration gradient (transport dissolved substances from regions of low to high concentration). This is accomplished through the action of compound-specific carriers (enzymes) located in the cell membrane that require the input of work (energy) to overcome the unfavorable transport of components to the higher concentration region inside the cell. It is this process (*active transport*) that allows organisms to thrive in nutrient-poor environments. The importance of energy conservation in the overall metabolism of microorganisms is evident, as the absence of an energy source would prevent the operation of active transport mechanisms for assimilation.

All of the fungi and algae and many bacteria are osmotrophic. However, molds and some heterotrophic bacteria (those incapable of synthesizing their own food and having to depend on an outside source) are recognized as important agents in the breakdown of particulate organic matter (*composting*). In these cases, the particles are solubilized by exocellular enzymes or metabolic products and the resulting dissolved organic matter is assimilated to the intracellular environment.

Ingestion. Two classes of protozoa, ciliates and amoebas, have the ability to ingest particulate organic matter. These protozoa are prevalent in wastewater treatment systems, where they feed principally on bacteria through a process known as predation. Phagotrophic protozoa also can consume algae, fungi, viruses, other protozoa, and *abiotic* (noncellular) particulate organic matter. These organisms must first solubilize the ingested particles intracellularly before the released organic molecules can contribute to the cell's nutrition. Intracellular solubilization is accomplished through the action of specific enzymes.

It is through intracellular solubilization that the metabolism of osmotrophic and phagotrophic organisms is unified: the *dissimilation* (breakdown) of organic compounds, biosynthesis, and energy conservation are then performed intracellularly through reactions involving dissolved organic substrates.

CELL MAINTENANCE. Cell maintenance refers collectively to the functions performed by the cell that are not directly associated with growth. That is, a certain amount of energy used by the cell is diverted to preservation of the status quo, which is necessary even in the absence of biosynthesis. In this way, cellular integrity is preserved. Failure to meet the requirements of cell maintenance will result in cell death.

Major functions involved in cell maintenance are *motility* (movement) and the preservation of structural integrity. Motile cells require mechanical energy for the operation of their locomotive processes. Means of cellular locomotion include the beating of flagella or cilia, and cytoplasmic streaming. As there exists in all organisms a natural tendency for macromolecules to spontaneously decompose, chemical energy is continually required for the resynthesis of structural components.

*M*ETABOLIC CLASSIFICATION

Differences in the details of metabolism provide an important means by which to classify organisms. Although there exist other ways by which cells can be identified and grouped, metabolic classifications are most relevant from a biochemical engineering perspective, as the chemical transformations that specific cells can *catalyze* (speed up) are of such primary interest. Metabolic classifications are made on the basis of carbon sources and energy sources.

CARBON AND ENERGY SOURCES. Because carbon typically composes a significant portion of the total dry mass of cells, all cells need an external source of carbon for biosynthesis. Only two such sources exist in nature, organic compounds and carbon dioxide. Organisms that can use carbon dioxide as their primary source of carbon are known as *autotrophs*. Many organisms, however, do not possess the ability to use carbon dioxide as a principal carbon source and must rely on organic compounds for this purpose. These organisms are known as *heterotrophs*. Cells may obtain energy either from organic or inorganic compounds assimilated from the extracellular environment or from harnessing energy present in visible light through specialized cellular apparati. Few organisms can use both energy sources. Those that use chemical compounds (organic or inorganic) for their primary energy source are known as *chemotrophs*, while those that extract energy from visible light are called *phototrophs*. Based on these terms, there is a classification scheme that divides organisms into four metabolic groups: chemoheterotrophs, chemoautotrophs, photoautotrophs, and photoheterotrophs.

Chemoheterotrophs. Chemoheterotrophs use organic compounds as their source of energy and rely on organic compounds as their primary carbon source. This group includes all fungi (yeasts and molds), protozoa, and many bacteria.

Chemoautotrophs. Chemoautotrophs use inorganic compounds for an energy source and can use carbon dioxide as the primary carbon source. Only certain highly specialized bacteria—including the nitrifying, hydrogen, sulfur, and iron bacteria—are chemoautotrophic.

Photoheterotrophs. Photoheterotrophs use light for a principal energy source and rely on organic compounds as their primary carbon source. Only a few algae and cyanobacteria are classified in this small group, which is dominated by anaerobic, photosynthetic bacteria.

Photoautotrophs. Photoautotrophs use light for an energy source and can use carbon dioxide as their primary carbon source. Most of the algae and the cyanobacteria (blue–green bacteria) belong to this group.

A general metabolic scheme for each of the four classes above is given in Figures 6.3 through 6.6. In chemoheterotrophs, the overall formation of assimilated organic molecules into central intermediates is an energy-yielding (ATP-synthesizing) process (Figure 6.3). This process of oxidation of organic compounds for the production of central intermediates with the simultaneous manufacture of ATP in chemoheterotrophs is called *catabolism*. Catabolic processes refer to destructive metabolic changes in a cell, involving the breakdown of compounds to simpler substances, yielding energy. By contrast, autotrophic fixation (reduction) of carbon dioxide to the central intermediates requires the expenditure of metabolic energy.

Thus, carbon assimilation and fixation add to the total load on the internal pool of metabolic energy in autotrophs. Essentially, photoautotrophs and

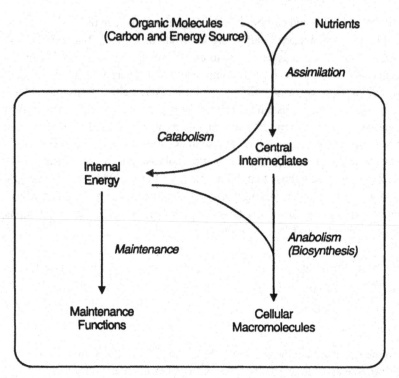

Figure 6.3   General overview of chemoheterotrophic metabolism.

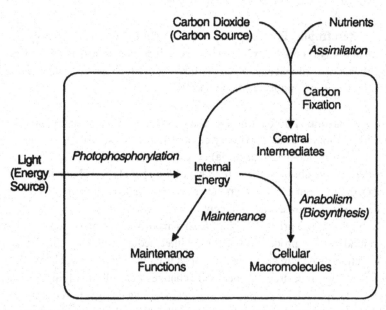

Figure 6.4   General overview of photoautotrophic metabolism.

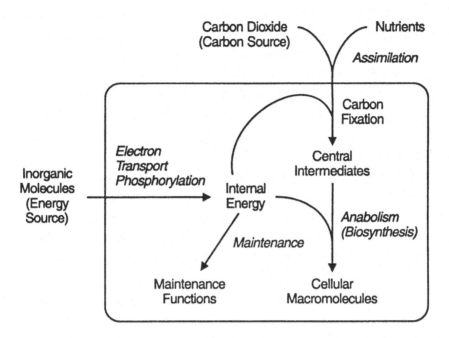

Figure 6.5   General overview of chemoautotrophic metabolism.

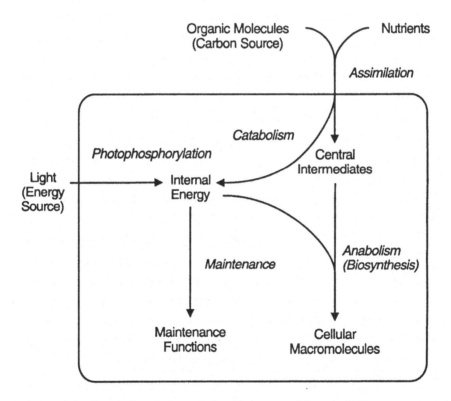

Figure 6.6   General overview of photoheterotrophic metabolism.

chemoautotrophs differ in the mechanisms by which ATP is formed from the external energy source. These mechanisms are termed *photophosphorylation*, meaning the production of ATP by photosynthesis (Figure 6.4) and electron transport phosphorylation (Figure 6.5), respectively. Photoheterotrophs produce ATP primarily through photophosphorylation (Figure 6.6), though a small fraction of the overall ATP may be derived from catabolism, as in the chemoheterotrophs.

NUTRIENT REQUIREMENTS

The assimilation of nutrients represents an important requirement for biosynthesis. Nutrients are the chemical components taken up by the cell from the external environment that are necessary for growth and maintenance. Nutrients may be classified as either major or minor bioelements. Major bioelements are those needed in high extracellular concentrations for normal metabolic operation. Some of these constitute the major portion of cell mass, while others are necessary for the activity of certain enzymes. Minor bioelements consist mostly of metal cations, which serve a variety of metabolic functions. Most of these functions are associated with the structural and catalytic properties of enzymes.

MAJOR BIOELEMENTS. There are 10 bioelements needed by the cell in high concentrations. Intact, live cells contain mostly water, which makes up approximately 70% of their weight. Thus, metabolism consists of reactions that occur in a principally aqueous environment. Most discussions of cell composition, however, focus on the nonaqueous constituents of cells, which collectively are the reactants, products, and catalysts of metabolism. The term *dry mass*, the residue of cells after heated evaporation or freeze drying, typically is used to represent this pool of nonaqueous components in cells. Based on elemental and macromolecular analyses of dried cells of the bacterium *Escherichia coli*, it is estimated that approximately 92% of cell dry mass is comprised of the bioelements carbon, hydrogen, oxygen, and nitrogen.

Cells obtain their carbon either through the heterotrophic oxidation of organic compounds, whereby metabolic energy (ATP) also is obtained, or by the autotrophic reduction of carbon dioxide, which requires the expenditure of metabolic energy. For autotrophic organisms, essentially all carbon obtained from the fixation of carbon dioxide is used for the production of cell mass (biosynthesis), as carbon does not contribute to their internal supply of energy. Heterotrophs, however, must distribute molecules of their carbon source(s) between both catabolic (energy producing) and anabolic (biosynthetic) pathways. Thus, part of the carbon assimilated in heterotrophic organisms is converted to cell mass, while the remaining fraction ends up as carbon dioxide or other products of energy conservation.

A small number of organisms, both autotrophic and heterotrophic, cannot synthesize all of the specific organic molecules necessary for growth, so the compounds must be present in (and assimilated from) the extracellular environment if biosynthetic reactions are to proceed. Certain vitamins and *amino acids* (the fundamental structural units of proteins) represent the most common requirements for preformed organic molecules.

Nitrogen, required by all microbes for the synthesis of proteins and nucleic acids—components of deoxyribonucleic acid (DNA) and ribonucleic acid (RNA)—can be obtained in various forms normally present in the environment. The preferred inorganic form is ammonia (NH_3), which requires some metabolic energy (but no net transfer of electrons) for its assimilation and transformation to cell material. Many, but not all, organisms also can assimilate nitrate (NO_3), which must be intracellularly reduced (accept electrons) to ammonia before its use in biosynthesis. This reaction is known as assimilatory nitrate reduction.

The extra metabolic energy required for the transfer of electrons from an internal source to the nitrate ion makes nitrate a less preferable nitrogen source than ammonia. Gaseous elemental nitrogen (N_2) also can be used by a limited number of specific organisms through an energy-consuming mechanism called nitrogen fixation, whereby the nitrogen atom again must be reduced to a compound such as ammonia before it can be used in biosynthesis.

Oxygen and hydrogen are present in cells principally as water but also compose a significant portion of the nonaqueous biomass, as they are basic components of all organic compounds. In heterotrophic organisms, which assimilate and transform organic compounds (substrate molecules) for carbon and energy, much of the oxygen and hydrogen incorporated to the major macromolecules comes from the substrate molecules themselves (such as glucose, $C_6H_{12}O_6$).

In autotrophs, the oxygen molecule from carbon dioxide is incorporated to cell material along with the carbon atom in the process of *carbon fixation.* Certain chemoheterotrophs possess enzymes, called *oxygenases*, that contribute to the direct incorporation of molecular oxygen ($1/2 \ O_2$) to cell material. That is, reduced organics, such as methane (CH_4) or benzene (C_6H_6), can be oxidized by the action of an oxygenase. The enzyme activates molecular oxygen and thereby permits the incorporation of oxygen to a more oxidized organic product, such as methanol (CH_3OH) or catechol [$C_6H_4(OH)_2$].

The primary function of molecular oxygen in chemoheterotrophs is to serve as a terminal electron acceptor in a process of ATP-generation referred to as (aerobic) *respiration.* This means that molecular oxygen is the final electron acceptor receiving hydrogen atoms and thus forming.

The remaining fraction of cell mass, approximately 8% of dry mass, is composed of the six other major bioelements: sulfur, phosphorus, potassium, magnesium, calcium, and iron.

Sulfur is a component of various amino acids and is required for the synthesis of proteins. Most organisms can satisfy their need for sulfur by reducing inorganic sulfate (SO_4^{2-}) in energy-consuming reactions similar to those for assimilatory nitrate reduction. Thus, this process is called assimilatory sulfate reduction. Certain bacteria, however, do not assimilate sulfate, but rather depend on the availability of reduced sulfur compounds (such as hydrogen sulfide) for incorporation to cellular organic material. Typically, these are the anaerobic heterotrophs.

Phosphorus is present in the form of a phosphate ion (PO_3^-) in several cellular organic compounds, such as nucleic acids (RNA, DNA), nucleotides (ATP), nucleosides (ADP, AMP), and phospholipids. Under certain conditions, some cells also store phosphorus in the form of *polyphosphates* (chains of phosphate molecules). Most organisms can assimilate and directly use orthophosphate ($H_3PO_4^-$), an inorganic form of phosphate.

While inorganic phosphate seems to be the major source of phosphorus for cell growth, organic phosphate compounds often occur in nature. These can be used by cells through the action of specific enzymes called *phosphatases*, which cleave the phosphate bond of the compound, thus releasing the phosphorus to participate in biosynthetic reactions. The primary mechanism for incorporation of phosphorus to cellular macromolecules is through the donation of a phosphate ion from ATP during anabolism. To sustain these reactions (thus, biosynthesis), assimilated phosphorus must be incorporated to ATP. Thus, ATP plays a central role as a phosphate carrier, along with its primary function as an energy carrier.

Potassium is required by all organisms and primarily functions in the activation of intracellular enzymes. Calcium is involved in the stability of the cell wall and is also a constituent of certain enzymes. Magnesium is involved in the activation of many enzymes and is also thought to act as a structural component of other cell constituents, such as chlorophyll molecules in photosynthetic cells. Potassium, calcium, and magnesium can all be assimilated as cations. Iron is required by all organisms and serves as a structural component of certain proteins. In organisms that create ATP through aerobic respiration (by using oxygen as a terminal electron acceptor), iron is a component of various carrier molecules that make up the electron transport chain, the primary apparatus responsible for ATP production. Iron can be assimilated in its ferric (+3 oxidation) state, or through its association with various organic compounds. The latter mechanism seems to be prevalent in environments with low concentrations of dissolved iron.

MINOR BIOELEMENTS. Several elements are required by organisms in minute quantities, and are termed the minor, or *trace*, bioelements. Zinc, manganese, cobalt, copper, and molybdenum are required by all organisms for various growth functions and play important roles in the activation and structural integrity of enzymes, energetic (energy conservation) pathways, and the formation of certain organic compounds, such as vitamins required for growth.

Some organisms require other trace bioelements, such as tungsten or nickel. A single group of algae, the diatoms, have a requirement for the element silicon, which is used as a structural component of the cell wall.

ENERGETIC REQUIREMENTS

The previous section primarily considered the uptake of nutrients required for biosynthesis. Equally essential is the process of energy conservation. The mechanism by which phosphorus participates in anabolic reactions is through the production of ATP and the subsequent donation of a phosphate ion from this molecule to an organic macromolecule. Thus, phosphorus is unique among the biosynthetic nutrients in that its assimilation for biosynthesis and its participation in energy conservation occur by the same reaction mechanism. Reactions of this type, which are distinguished by a phosphate ion transfer, are known as phosphorylations. This phenomenon may occur by substrate-level phosphorylation, electron transport phosphorylation, or photophosphorylation.

The conservation of energy in biological systems requires the transfer of electrons between donor and acceptor molecules, with the energy source being the electron donor. In chemoheterotrophs, for example, the primary energy source is an organic molecule, and the donated electrons typically are initially transferred to NAD, which would then be referred to as the primary electron acceptor. Thus, the ability of the donor molecule (organic substrate) to donate electrons depends on the availability of an adequate intracellular supply of oxidized primary acceptor (NAD). Because only a finite amount of NAD is produced in the cell, it must be *cycled* (reduced and subsequently oxidized), so that a primary acceptor molecule is always available for the oxidation of an external energy (electron) source. The ultimate fate of the electrons given up by the NADH molecule will determine the amount of metabolic energy (ATP) obtained and is characteristic of the particular energetic (phosphorylation) mechanism by which these ATP-generating reactions are carried out.

The molecule or ion that ultimately receives the electron in energy conservation reactions is called the *terminal electron acceptor*. Examples of terminal electron acceptors include molecular oxygen in organisms that conduct aerobic respiration, nitrate and sulfate in organisms that respire anaerobically (without oxygen), and an organic molecule in fermentative cells (yeast, which do not require oxygen for synthesis).

The amount of ATP that can be generated by oxidation-reduction depends on the number of electrons transferred. Thus, the potential amount of energy production through a particular mechanism can be determined by the difference between the number of electrons associated with the donor molecule and the number associated with the oxidized product formed by the electron donation. This difference in oxidation states between the donor and product indicates the number of electrons available for transfer to the terminal acceptor, thus the amount of ATP that can theoretically be produced. The primary difference among the three mechanisms is the molecule used as the terminal acceptor and the means by which the electron transfer is accomplished.

*P*ERTINENCE TO WASTEWATER TREATMENT

The principal objectives of various biological systems for wastewater treatment are carbon oxidation, nitrification, and denitrification. These objectives and their rationales can be related to features of microbial metabolism that already have been described.

RATIONALES. Municipal and industrial liquid wastes typically contain numerous, soluble organic compounds that can serve as carbon sources for osmotrophic chemoheterotrophs, and insoluble organic matter that can be solubilized to dissolved organic matter. Upon indiscriminate disposal of untreated wastes, the activity of aerobically respiring chemoheterotrophs native to the receiving environment can be stimulated. Substantial depletion of oxygen in the discharge zone can be an immediate result of such activity, known as a carbonaceous biological oxygen demand. To avoid such oxygen depletion in the receiving stream, it is necessary to remove organic matter from the waste stream.

For soluble organics, removal requires transformations that render the carbon less harmful. This can be accomplished by oxidizing the compound to a point where it is no longer able to serve as an electron donor (carbon dioxide) through a process called mineralization, by producing gases (methane or carbon dioxide) that can be stripped from solution, or by producing new cells (*biomass*) that can be separated from the liquid effluent by sedimentation and filtration. Because the carbon present in cell mass exists in a more reduced form (as sugars, proteins, and lipids) than carbon dioxide, it can serve as an electron donor and still represents an oxygen demand. To the extent that the formation of carbon dioxide is maximized at the expense of biomass formation, the overall extent of mineralization (with a simultaneous reduction in the total potential oxygen demand) is increased. Thus, the transformation (oxidation) of organic matter to carbon dioxide is a worthy treatment objective.

For insoluble organics, such as fats, waxes, starch, and large protein molecules, removal requires *digestion* (the exocellular dissolution and breakup of particulate organic matter). Transformation of the soluble organics released by digestion proceeds as described above.

Ammonia-nitrogen and nitrite-nitrogen in municipal and industrial wastes also can stimulate oxygen depletion in the discharge zone if they are present in discharge effluents. This is referred to as a nitrogenous biological oxygen demand, and is caused by the activity of chemoautotrophic, ammonium- and nitrite-oxidizing bacteria native to the receiving environment. Thus, oxidation of inorganic nitrogen (nitrification) often is established as a treatment objective.

As nutrients for cellular synthesis, nitrogen—whether as ammonia, nitrite, urea, or nitrate—and phosphorus can stimulate the production of excessive amounts of algae and cyanobacteria (algal blooms) in lakes, ponds, and reservoirs. This is referred to as *eutrophication* (nutrient enrichment of aquatic systems). It is to be avoided, as such growths can result in aesthetically unpleasant color, taste, and odor being imparted to the water and can settle to deeper water, where they will respire and decompose (thereby instigating oxygen depletion in a manner similar to wastewater organics).

When these effects of nitrogen are of concern, removal of all forms of nitrogen is necessary. This can be accomplished by performing nitrification and denitrification. When phosphorus is of concern as a potential stimulant of eutrophication, it has been commonplace in the past to remove it through chemical precipitation. Increasingly, biological means of phosphorus removal have been used in recent years to accomplish this objective.

The detailed mechanisms of biological phosphorus removal are still the subject of some debate, but it is generally accepted that sequentially exposing certain chemoheterotrophs to anaerobic, and then to aerobic, conditions can result in the overproduction of phosphorus-rich macromolecules in the cells. In this way, soluble phosphorus in the waste is incorporated to biomass, which can be separated from the liquid effluent by physical means.

CARBON OXIDATION. The most common manner for performing carbon oxidation is by using aerobic treatment systems, such as activated sludge, trickling filters, aerated lagoons, and oxidation ponds. These systems can differ with respect to the processes for oxygenation, but all rely on the metabolism of aerobically respiring chemoheterotrophs to mineralize organic matter (oxidize organic matter to innocuous carbon dioxide).

Carbon oxidation also can be accomplished anaerobically. In anaerobic systems, carbon dioxide is formed through the concerted activity of fermenting bacteria, sulfate-respiring bacteria, and certain methanogenic bacteria. Another feature of anaerobic treatment systems is that some of the organic carbon also is reduced. This will occur in all anaerobic systems because of the metabolism of fermenting bacteria.

In extremely anaerobic systems, overall carbon reduction can result in methane formation. When methane, which is sparingly soluble, is stripped from the waste, it represents a removal of reduced organic carbon. When the waste derived methane is subsequently burned for energy, carbon dioxide is a primary product of combustion. Thus, the removed carbon is transformed to its most oxidized (thus most innocuous) state. Methanogenesis occurs through either reduction of the methyl group (CH_3) of acetic acid (CH_3COOH) or reduction of carbon dioxide. Both of these processes participate in the cell's energy metabolism. It is currently believed that both pathways involve an electron-transport phosphorylation to produce ATP and methane.

Carbon oxidation also will occur in systems for denitrification, which are *anoxic* (lack oxygen). With terminal (or tertiary) denitrification, a carbon source such as methanol (CH_3OH) is added to sustain the heterotrophic metabolism of bacteria that can respire with nitrate as the terminal electron acceptor in the absence of oxygen. As with aerobic respiration, organic carbon is converted to carbon dioxide and cellular macromolecules (biomass).

NITROGEN REMOVAL. *Nitrification* is the term given to the overall reaction in which ammonia-nitrogen (NH_3) is oxidized to nitrite-nitrogen (NO_2), which is itself oxidized to nitrate-nitrogen (NO_3). Each step in this process of nitrogen oxidation is catalyzed by a different group of specialized, chemoautotrophic bacteria that respire aerobically. The metabolism of the nitrite-oxidizing bacteria differs only in the nature of the electron donor (nitrite instead of ammonia) and the product of electron donation, or oxidation (nitrate instead of nitrite). Because nitrification involves aerobically respiring (chemoautotrophic) bacteria, it is an objective that is compatible with any aerobic system for carbon oxidation.

Denitrification, the reduction of nitrate to molecular gaseous nitrogen (N_2), is catalyzed by a diverse but select group of anaerobically respiring bacteria. Under conditions of negligible or no dissolved oxygen, these bacteria can use nitrate as the terminal electron acceptor to produce ATP through electron transport phosphorylation. The initial product is nitrite. However, under appropriate pH and other environmental conditions, additional reduction reactions can occur to yield nitrogen as the ultimate product of respiratory metabolism.

Not all denitrifying bacteria are heterotrophic; autotrophic (sulfur-oxidizing or hydrogen-oxidizing) denitrifiers also exist. However, heterotrophic bacteria typically are emphasized to accomplish this objective. Increasingly, denitrification is being applied as a primary rather than terminal biological process. In this way, wastewater organics, rather than purchased methanol, are used to sustain heterotrophic metabolism. Because carbon oxidation still occurs anaerobically, the beneficial result is a reduced organic loading for the aerobic stage.

BIOSYNTHESIS. An unavoidable consequence of all biological processes for wastewater treatment is the production of new cells, also referred to as a sludge, composed of wet biomass. Actively growing cells are relied on for the necessary transformations in conventional biological treatment systems. Consequently, attaining the desired extent of treatment necessitates that all required nutrients be present and appropriate environmental conditions be maintained for growth. In this regard, it may be necessary to supplement certain industrial wastes—which can be nutrient deficient—with nitrogen, phosphorus, or trace metals to bring about substantial carbon oxidation.

Biomass produced during chemotrophic processes for carbon, nitrogen, and phosphorus removal must be separated from the liquid effluent for maximum treatment benefits. Aerobically respiring biomass, even in the absence of available chemical substrates, can consume oxygen by respiration, which supplies maintenance energy. Moreover, biomass also represents a form of particulate organic matter that, on decomposition and solubilization, can release soluble organics that can impart an oxygen demand (just as does noncellular organic particulate matter in raw wastes). Therefore, it is necessary to not only separate biomass from liquid effluent but also to stabilize the biological sludge, ideally by some process for carbon oxidation (aerobic or anaerobic digestion).

For autotrophs, the process of carbon fixation leads to synthesis of biomass. This can proceed readily in the absence of measurable quantities of dissolved organic matter, as the cells use inorganic carbon dioxide as their carbon source. Consequently, it is not uncommon in oxidation ponds and facultative lagoons (where photoautotrophic algae and cyanobacteria are relied on to produce oxygen through noncyclic photophosphorylation) that the effluent can contain higher levels of volatile suspended solids than raw influent wastewater. This is attributable to extensive fixation of carbon by photoautotrophs. As with chemotrophic biomass, if biomass is not removed from the final effluent—as through settling, screening, or filtration—a substantial oxygen demand can be exerted in the receiving stream.

REFERENCE

Brock, T.D., *et al.* (1984) *Biology of Microorganisms*. 4th Ed., Prentice Hall, Inc., Englewood Cliffs, N.J.

SUGGESTED READINGS

Bailey, J.E., and Ollis, D.F. (1986) Biochemical Engineering Fundamentals. 2nd Ed., McGraw-Hill, Inc., New York, N.Y.

Comeau, Y., *et al.* (1986) Biochemical Model for Enhanced Biological Phosphorus Removal. *Water Res.*, **20**, 12, 1511.

Gaudy, A.F., Jr., and Gaudy, E.T. (1980) *Microbiology for Environmental Scientists and Engineers*. McGraw-Hill, Inc., New York, N.Y.

Grady, C.P.L., Jr., and Lim, H.C. (1980) *Biological Wastewater Treatment. Theory and Applications*. Marcel Dekker, New York, N.Y.

Large, P.J. (1983) Methylotrophy and Methanogenesis. In *Aspects of Microbiology*. Volume 8, Am. Soc. Microbiol., Washington, D.C.

Chapter 7
Heavy Metals

Studies relating to the chemistry, treatment, biological effects, environmental fate, and control of metals in the aquatic environment have been given increased attention recently. This is primarily because of the recognition of potential adverse health and environmental effects of metals discharged to the environment from municipal and industrial wastewater treatment facilities through either wastewater effluents or sludge solids. Metals commonly found in municipal effluent and sludge include cadmium, copper, lead, mercury, nickel, and zinc.

In wastewater, the presence of high concentrations of heavy metals is a serious concern because they can

- Inhibit the biological activity in secondary treatment systems, resulting in high levels of effluent organics;
- Deteriorate the biological activity in secondary treatment systems, resulting in high levels of effluent organics;
- Inhibit the biological nitrification process;
- Accumulate on sludge solids and prevent their disposal by incineration or land application;
- Inhibit the anaerobic sludge digestion process;

- Cause adverse environmental effects when discharged to surface waters through wastewater effluents; and
- Cause sludges to be regulatorily or agronomically unusable for land disposal.

HEAVY METALS IN MUNICIPAL WASTEWATER INFLUENTS AND SLUDGES

Metals found in municipal wastewater originate from a variety of industrial, commercial, and residential activities and from atmospheric deposition and stormwater runoff (Davis and Jacknow, 1975, and Kodukula and Obayashi, 1979). The contribution of heavy metals from residential and industrial sources is affected by the number and nature of contributing industries and pretreatment regulations. Acidity of water supply and presence of combined sewers can also affect effluent metals concentrations.

Whereas no strong correlation has been found between the percent of industrial flow and metal concentrations in influents to wastewater treatment plants (WWTPs), most cities with less than 4% industrial contribution have lower influent metals concentrations in their waste discharges (Minear *et al.*, 1981).

Data on the concentrations of selected metals in raw wastewater from a survey of 90 WWTPs across the country are presented in Table 7.1. The results indicate that the concentration ranges vary greatly, being related to fluctuating industrial outputs and unpredictable stormwater runoff. This is probably caused by treatment systems receiving almost zero to more than 50% industrial flow.

Table 7.1 Range and average metals concentrations in wastewater treatment plant influent (AMSA, 1989).*

Metal	Range, mg/L	Average, mg/L
Arsenic	0.000 1 – 5	0.187
Chromium	0.000 6 – 9	0.255
Copper	0.004 – 20	0.637
Lead	0.001 – 5	0.138
Mercury	0.000 1 – 0.5	0.034
Nickel	0.002 – 7	0.294
Silver	0.000 2 – 5	0.173
Zinc	0.024 – 2 027	39.4

* Based on survey results from 90 wastewater treatment plants.

*H*EAVY METAL REMOVAL IN *ACTIVATED SLUDGE SYSTEMS*

Heavy metals entering wastewater treatment facilities either remain in the wastewater stream and are discharged to receiving waters through the effluent, or are removed through the primary and secondary sludge solids. Therefore, the removal of metals from wastewater influents is a function of both their ability to partition onto primary and secondary solids and the efficiency of the treatment process in removing solids.

PRIMARY SEDIMENTATION. Metal removal during primary sedimentation is a physical process, depending on the settling of metals associated with the settleable solids of raw wastewater and on solids removal efficiency of the sedimentation process. Settleable precipitates of metals, if any, also are removed.

A summary of influent metal concentrations to primary sedimentation units and average removal efficiencies for several treatment plants in major metropolitan areas is presented in Table 7.2. The percentage of removals are not constant for all metals in any given system or for any given metal in all systems. Furthermore, the influent concentration of a given metal and its percent removal are not related. The variability in metal removal in primary treatment can be caused by differences in either the solids removal efficiency of the clarifiers, or the physical and chemical characteristics of the influents. Table 7.2 shows that cadmium is the least removed metal, while chromium, copper, lead, nickel, and zinc exhibited the highest removals during primary treatment.

ACTIVATED SLUDGE. Heavy metals in primary effluents entering activated sludge units are either removed through the secondary sludge or remain

Table 7.2 Summary of influent metal concentrations and percent removals of primary treatment in selected municipal wastewater treatment plants.

Location	Cadmium	Chromium	Copper	Lead	Nickel	Zinc
Chicago, Illinois	0.029	0.180	0.053	0.180	0.050	0.600
(U.S. EPA, 1982)	(10%)	(17%)	(8%)	(17%)	(60%)	(15%)
Pittsburgh, Pennsylvania	0.021	0.095	0.127	0.119	0.078	0.648
(Minear *et al.*, 1981)	(14%)	(17%)	(23%)	(54%)	—	(19%)
Four different cities in	0.021	0.299	0.143	0.228	—	0.615
U.S. (McCalla *et al.*, 1977)	(14%)	(27%)	(50%)	(42%)	—	(53%)
1–12 different cities in	—	—	—	—	—	—
U.S. (Blakeslee, 1973)	(15%)	(27%)	(22%)	(57%)	(14%)	(29%)

in the effluent. The degree of removal depends on the partitioning of the metals onto the activated sludge biomass and its subsequent removal in the secondary clarifiers. The average concentrations of selected metals in secondary influents and effluents of full-scale activated sludge systems in selected major U.S. cities are given in Table 7.3. There is a general decrease of metal concentrations from raw wastewater to primary effluent to secondary effluent, indicating accumulation of a portion of the metal onto the solids and their subsequent removal. Table 7.3 shows metal removal efficiencies during activated sludge treatment to be in the order of Zn > Cd > Pb > Cr > Cu > Ni.

Table 7.3 Summary of influent and effluent concentrations of metals in selected activated sludge treatment plants.

Location		Cadmium	Chromium	Copper	Lead	Nickel	Zinc	Reference
Bryan, Ohio	Influent	—	0.8	0.2	—	0.05	2.2	Doty et al.,
	Effluent	—	0.2	0.1	—	0.05	0.2	1977
Grand Island,	Influent	0.018	0.059	0.17	0.16	—	0.353	McCalla
Nebraska	Effluent	0.016	0.013	0.067	0.092	—	0.182	et al., 1977
Grand Rapids,	Influent	—	3.6	1.4	—	2.0	1.5	Doty et al.,
Michigan	Effluent	—	2.5	1.6	—	1.8	0.8	1977
Hyperion,	Influent	0.028	0.3	0.13	0.11	0.2	0.43	Konrad and
California	Effluent	0.028	0.21	0.13	0.10	0.14	0.26	Kleinert, 1974
Muncie,	Influent	—	0.26	0.26	0.93	0.13	0.97	Minear
Indiana	Effluent	—	0.05	0.07	0.22	0.11	0.26	et al., 1981
New York,	Influent	0.016	0.16	0.27	—	0.11	0.41	Council,
New York	Effluent	0.01	0.08	0.15	—	0.10	0.21	1976
Rockford,	Influent	0.25	—	1.17	—	0.37	2.8	Furr et al.,
Illinois	Effluent	0.05	—	0.19	—	0.32	0.45	1976
Average	Influent	0.078	0.863	0.514	0.40	0.477	1.238	
concentration	Effluent	0.026	0.509	0.330	0.137	0.42	0.337	
Average removals, %		67	41	36	66	12	73	

*T*OXICITY OF HEAVY METALS

Although trace concentrations of heavy metals are needed for biological growth, high concentrations can be inhibitory or toxic. The toxicity of heavy metals in biological systems is believed to be caused by the free metal ions in the soluble phase. The metal ions can form bonds or inactive complexes with unused electrons at the active sites of many enzymes, suppressing substrate uptake and use and other metabolic processes.

The inhibitory threshold levels of heavy metals for biochemical oxygen demand (BOD) removal and nitrification in activated sludge systems are presented in Table 7.4. It appears from these data that arsenic, mercury, and zinc inhibit BOD removal; whereas copper, zinc, and chromium have more effect on nitrification.

There are several ways to detect inhibitory and toxic effects of heavy metals in activated sludge systems. Early detection of such effects is possible by observing the following indicators:

- A decrease in the protozoan activity;
- The disappearance of protozoa;
- A decrease in the oxygen uptake rate;
- An increase in the nitrite concentration;
- Deterioration of effluent quality, especially with regard to effluent solids; and
- An increase in lethality during on-line bioassay testing.

Although inhibitory or toxic effects of metals can be indicated by the above indicators, they do not necessarily prove metal toxicity.

Inhibitory and toxic effects of metals also have been observed in anaerobic sludge digestion systems. Table 7.4 also includes toxic levels of metals in anaerobic systems. Based on the data in Table 7.4, one might conclude that the metals are more toxic to activated sludge biomass than to anaerobically digesting sludge, as the threshold values are lower for the former. However,

Table 7.4   Reported range of inhibition threshold levels for selected heavy metals in activated sludge nitrification systems, and anaerobic digesters (Barth *et al.*, 1965).

Metal	Activated sludge, mg/L	Nitrification systems, mg/L	Anaerobic digesters, mg/L
Arsenic	0.1	1.5	1.6
Cadmium	1 – 10	5.2	20
Chromium (total)	1 – 100	0.25 – 1.9	—
Chromium (III)	10 – 50	—	130
Chromium (VI)	1	1 – 10	110
Copper	1	0.05 – 0.48	40
Lead	0.1 – 100	0.5	340
Mercury	0.1 – 1.0	—	—
Nickel	1.0 – 5.0	0.25 – 5.0	10 – 130
Silver	0.25 – 5.0	—	13 – 65*
Zinc	0.08 – 10	0.08 – 0.5	400

* Dissolved metal.

toxicity is not simply a function of toxic metal concentration alone, but also to the metals-to-biomass ratio. An explanation for lower inhibitory values for activated sludge may arise from the lower biomass concentration typical of activated sludge systems when compared to anaerobic digesters.

Studies reporting metal toxicity levels in biological treatment processes and receiving waters typically do not specify the form of the metal (whether it is the soluble or total metal). Therefore, the toxic concentrations should be total concentrations rather than soluble.

DISTRIBUTION OF METALS IN WASTEWATER STREAMS

Heavy metals in wastewater streams are distributed between the solid and liquid phases. From an operational standpoint, the portion of the metal associated with the solids retained by a 0.45-µm filter is classified as solid-phase metal. The soluble portion is what passes through the filter. In the soluble phase, a given metal can be present as a free ion or as a soluble complex. In the solid phase, it can be either in a precipitate form, associated with raw wastewater solids, or the activated sludge biomass. These different metal species are in equilibrium with each other, as depicted in Figure 7.1. The relative distribution of metals between the soluble and solid phases is a function of

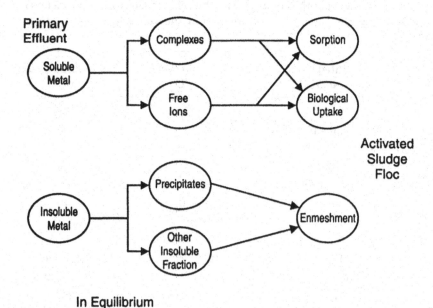

Figure 7.1 Metal interactions in activated sludge.

different mechanisms, including complexation in soluble phase, precipitation, sorption, and biological uptake.

COMPLEXATION IN SOLUBLE PHASE. *Complexation* is a process whereby a positively charged metal ion attaches or bonds to a molecule or a charged ion, which is called a *ligand* or a complexing agent. Such agents form stable soluble metal complexes, keeping those metals in solution that might otherwise precipitate or accumulate on the sludge solids. This results in higher effluent concentrations and lower metal-removal efficiencies.

Complexing agents commonly found in wastewater can be either inorganic or organic. Common inorganic complexing agents include carbonate, hydroxide, sulfate, chloride, phosphate, and ammonia. Organic complexing agents can be either natural (humic and fulvic acids) or synthetic (Nitrilo-triacetic acid [NTA] and Ethylenediaminetetraacetic acid [EDTA]). Organic complexes are much stronger than the inorganic ones, resulting in significantly lower metal removals in wastewater treatment processes. Whereas strong, synthetic organic complexing agents are not normally found in municipal wastewater, naturally occurring organics are common.

PRECIPITATION. *Precipitation* of a metal occurs when its solubility limit is reached. The solubility of a metal in an aqueous solution is primarily a

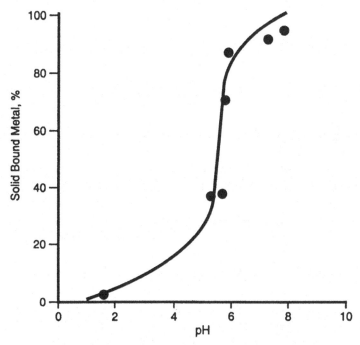

Figure 7.2   Effect of pH on cadmium adsorption in activated sludge biomass (total suspended solids = 2 600 mg/L; total cadmium = 0.8 mg/L) (Kodukula, 1984).

function of pH and competing ions. As the pH decreases, solubility increases. The solubility of metals also is influenced by temperature and complexing agents. As temperature of the medium increases, the metal solubility also increases.

SORPTION. *Sorption* represents the physical and chemical interactions between metals and particulate matter in the wastewater process streams. Particulate matter typically includes raw wastewater solids, inorganic debris (for example, clay), biomass (activated sludge floc), and colloidal organic material (such as protein, carbohydrate, lipid, and fatty acid substances).

Three major processes of sorption take place on the surface of the particulate matter: adsorption, complexation, and ion exchange. Adsorption of metal onto the solid surface involves a chemical reaction or a physical process. In the first case, chemisorption, the adsorbed ion undergoes chemical interaction (for example, covalent bonding) with the adsorbent; whereas physical adsorption occurs when weak physical forces (Van der Waals forces) of the adsorbed molecule prevent it from being fixed to a specific binding site (Weber, 1972).

The complexation mechanism discussed earlier also can result in an attachment or bonding of cationic metal ions to a molecule or a charged ion in the solid phase. *Ion exchange* is a process in which ions held by electrostatic forces to charged functional groups on the surface of a solid are exchanged for similar ions.

Because it is difficult to distinguish among the above processes, it is common practice in the literature to consider these processes as a generic mechanism called sorption; the terms sorption and adsorption are used interchangeably.

Metal sorption reactions are believed to occur quickly. If a metal solution is added to activated sludge, metal cations are sorbed onto the solids in a rapid first phase, followed by a prolonged second stage (Cheng *et al.*, 1975). Most of the soluble metal is removed during the first stage, while the second stage seems to be insignificant. The biological floc of the activated sludge particles and the extracellular polymers (such as capsules or slime, or glycocalyx) produced by the bacteria play a key role in sorption of heavy metals (Brown and Lester, 1979). The organisms, either single cells or aggregates of cells present in the biological floc, are electronegative in the operational pH range of the activated sludge process. The surface charge of the organisms is a result of the ionization of some of the functional groups of the floc particles.

BIOLOGICAL UPTAKE. Metal uptake by biological sludge is demonstrated to be predominantly a physical and chemical process with biological uptake being insignificant (Nelson *et al.*, 1981). It has been reported that in media containing bacterial cultures, lead, copper, and most of the metal removed were found in the cell wall or extracellular polymers, with only negligible amounts in the cytoplasm.

FACTORS AFFECTING DISTRIBUTION OF METALS

Distribution of metals between the operationally defined soluble and solid phases of activated sludge is a complex function of physical and chemical characteristics of the wastewater streams, such as pH and concentration of solids, complexing ligands, and concentrations of the individual metal. It also depends on operational factors, including solids retention time (SRT) of the system. These factors determine the form in which the metal exists and what distribution mechanism is predominant under given conditions, which, in turn, control the metal partitioning behavior.

PHYSICAL AND CHEMICAL CHARACTERISTICS. The most significant physical and chemical factors influencing metal-partitioning behavior are pH, concentration of suspended solids and total metal, and strength and concentration of complexing ligands. The distribution of metals between the soluble and solid phases of solid and liquid suspensions is largely controlled by pH. It has been common practice to study the sorption behavior of metal cations by relating percent sorbed metal with the pH of the test liquid at fixed total metal and solids concentrations. Such a relationship has been shown to follow steep S-shaped curves, referred to as pH adsorption edges. An example of such a curve is presented in Figure 7.2.

Several investigators have reported that heavy metal sorption to activated sludge solids increases from approximately 0 to almost 100% over a narrow pH range of two units. It appears that for most cationic metals at pH levels greater than approximately 7.5, more than 80% of the metal is associated with the solid phase. Activated sludge mass has tremendous metal sorption capacity and resiliency relative to metals concentrations typically found in wastewater treatment plants, and the sorption capacity for cadmium and nickel increases logarithmically with pH (Kodukula, 1984).

Metal-partitioning behavior in activated sludge systems is significantly affected by aeration solids (MLSS) concentration. The fraction of sorbed metal has been observed to increase with increased aeration solids concentrations at a fixed pH and total metal concentration. Along with the concentration of MLSS, their size distribution strongly influences metal partitioning. Evidence as to whether certain solids size fractions are more favorable for sorption of specific metals indicates that the metals are unevenly distributed among different solids size fractions and between settleable and nonsettleable solids (Patterson and Kodukula, 1984).

The partition of a given metal between the soluble and solid phases and its speciation in the soluble phase of activated sludge are influenced by the presence of charged complexing ligands. The concentration and strength of the complexing agents are important. Inorganic agents such as chloride,

---

ammonia, and phosphate have little effect on metal distribution, whereas the influence of soluble organic carbon, and particularly of chelating agents such as NTA (Rossin *et al.*, 1982) and EDTA (Kao *et al.*, 1982), is significant. This has been demonstrated by a strong correlation between effluent BOD_5 and the effluent metals of several wastewater treatment plants in the U.S. (California, Illinois, and New York) (Patterson, 1978).

OPERATIONAL FACTORS. The single most important operational parameter that has been shown to influence metal partitioning behavior is the SRT. There have been several reports showing increased removal of metals at higher SRTs in continuously operated bench-scale simulations of activated sludge processes (Farroq and Gabriel, 1983; Sterrit and Lester, 1981; and Stoveland and Lester, 1980). Metal removal is enhanced at higher SRTs in at least two different ways. Under fixed conditions, increasing the SRT of an activated sludge system would result in an increase in MLSS concentration and a decrease in soluble organics. Also, with increasing SRT, the characteristics of the biomass (and as a result, sorptive properties of the solids) are altered. For example, it has been shown that production of extracellular polymers is enhanced at higher SRTs (Brown and Lester, 1979), which increases metal sorption.

*R*EFERENCES

Association of Metropolitan Sewerage Agencies (1989) Survey. Washington D.C.

Barth, E.F., *et al.* (1965) Summary Report on the Effects of Heavy Metals on the Biological Treatment Processes. *J. Water Pollut. Control Fed.*, 37, 86.

Blakeslee, P.A. (1973) Monitoring Considerations for Municipal Wastewater Effluents and Sludge Application to the Land. In *Recycling Municipal Sludge and Effluents on Land.* Natl. Assoc. State Univ. and Land Grant Coll., Washington, D.C.

Brown, M.J., and Lester, J.N. (1979) Metal Removal in Activated Sludge: The Role of Bacterial Extracellular Polymers. *Water Res.*, 13, 817.

Cheng, M.H., *et al.* (1975) Heavymetal Uptake by Activated Sludge. *J. Water Pollut. Control Fed.*, 47, 362.

Council for Agricultural Science and Technology (1976) Application of Sewage Sludge to Cropland: Appraisal of Potential Hazards of the Heavy Metals to Plants and Animals. Rep. No. 64.

Davis, J.A., and Jacknow, T. (1975) Heavy Metals in Wastewater in Three Urban Areas. *J. Water Pollut. Control Fed.*, 47, 2292.

Doty, W.T., *et al.* (1977) Chemical Monitoring of Sewage Sludge in Pennsylvania. *J. Environ. Qual.*, 6, 421.

Farroq, S., and Gabriel, P.F. (1983) Removal of Nutrients and Heavy Metals from Anaerobic Waste by Aerobiological Treatment. *Environ. Int.*, **9**, 69.

Furr, A.K., *et al.* (1976) Multielement and Chlorinated Hydrocarbon Analyses of Municipal Sewage Sludges of American Cities. *Environ. Sci. Technol.*, **10**, 683.

Kao, J.F., *et al.* (1982) Effects of EDTA on Cadmium in Activated Sludge Systems. *J. Water Pollut. Control Fed.*, **54**, 1118.

Kodukula, P.S. (1984) *Distribution of Cadmium and Nickel in Activated Sludge Systems*. Ph.D. thesis, Ill. Inst. Technol., Chicago.

Kodukula, P.S., and Obayashi, A.W. (1979) Accumulation of Toxic Priority Pollutants in Municipal Sludges. *Proc. Specialty Conf. on Control of Specific (Toxic) Pollut.*, Air Pollut. Control Assoc.

Konrad, J.G., and Kleinert, S.J. (1974) Removal of Metals from Wastewaters by Municipal Sewage Treatment Plants. In *Surveys of Toxic Metals in Wisconsin*. Dep. of Nat. Resources, Tech. Bull. 74, Madison, Wis.

McCalla, T.M., *et al.* (1977) Properties of Agricultural and Industrial Wastes. In *Soils for Management of Organic Waters and Wastewaters*. ASA-CSSA-SSSA, Madison, Wis., **11**.

Minear, R.A., *et al.* (1981) Database for Influent Heavy Metals in Publicly Owned Treatment Works. Rep. to Munic. Environ. Res. Lab., U.S. EPA.

Nelson, P.O., *et al.* (1981) Factors Affecting the Fate of Heavy Metals in the Activated Sludge Process. *J. Water Pollut. Control Fed.*, **53**, 1323.

Patterson, J.W. (1978) Heavy Metals Removal in Combined Wastewater Treatment. Paper Presented at Int. Environ. Colloquium, Univ. of Liege, Belg.

Patterson, J.W., and Kodukula, P.S. (1984) Metals Distributions in Activated Sludge Systems. *J. Water Pollut. Control Fed.*, **56**, 432.

Rossin, A.C., *et al.* (1982) The Removal of Nitrilotriacetic Acid and its Effects on Metal Removing During Biological Sewage Treatment: Part 1: Adsorption and Acclimation. *Environ. Pollut. (Ser. A)*, **29**, 271.

Sterrit, R.M., and Lester, J.N. (1981) The Influence of Sludge Age on Heavy Metals Removal in the Activated Sludge Process. *Water Res.*, **15**, 59.

Stoveland, S., and Lester, J.N. (1980) A Study of the Factors Which Influence Metal Removal in the Activated Sludge Process. *Sci. Total Environ.*, **16**, 37.

U.S. EPA (1982) *Fate of Priority Pollutants in Publicly Owned Treatment Works*. EPA/44/1-82/303, Washington, D.C.

Weber, W.J., Jr. (1972) *Physico-Chemical Processes for Water Quality Control*. Wiley-Interscience, New York, N.Y.

SUGGESTED READINGS

Beers, A. (1979) *Heavy Metals Equilibria in Activated Sludge Process*. M.S. thesis, Ill. Inst. Technol., Chicago.

Bogusch, E. (1974) Monitoring Study of Metals at the Calumet Treatment Works. Unpublished Report. Metropolitan Sanit. Dist. of Greater Chicago, Ill.

Brown, N.G., *et al.* (1973) Efficiency of Heavy Metals Removal in Municipal Sewage Treatment Plants. Environ. Lett., **5**, 103.

Chen, K., *et al.* (1974) Trace Metals in Wastewater Effluents. *J. Water Pollut. Control Fed.*, **46**, 2663.

Corpe, W.A. (1964) Factors Influencing Growth and Polysaccharide Formation by Strains of *Chromobacterium violaceum. J. Bacteriol.*, **88**, 1433.

Klein, L.A., *et al.* (1974) Sources of Metals in New York City Wastewater. *J. Water Pollut. Control Fed.*, **46**, 2653.

Lankford, P.W. (1991) Technologies for Toxicity Removal at POTWs. Paper presented at U.S. EPA Munic. Technol. Transfer Workshop, Kansas City, Mo.

McKinney, R.E. (1956) Biological Flocculation. In *Biological Treatment of Sewage and Industrial Wastes*, Vol. 1, Reinhold Publ. Corp., New York, N.Y.

Patterson, J.W., and Kodukula, P.S. (1978) Heavy Metals in Great Lakes. *Water Qual. Bull.*, **3**, 4.

Sposito, G., *et al.* (1977) Proton Binding in Fulvic Acid Extracted from Sewage Sludge-Soil Mixtures. *Soil Sci. Soc. Am. J.*, **41**, 1119.

Tornabene, T.G., and Edwards, H.H. (1972) Microbial Uptake of Lead. *Science*, **176**, 1334.

Chapter 8
Organic Compounds

A primary goal of biological wastewater treatment is to remove organic compounds from wastewater. Organic compounds are used by bacterial cells as electron donors or acceptors in energy producing reactions, as carbon sources for cellular synthesis, or as required nutrients or cofactors for bacterial

---

metabolism. During biological treatment, organic compounds are converted to gaseous, particulate, and soluble products. Gases such as carbon dioxide or methane are removed from the liquid phase of the wastewater by transfer to the gas phase. Particulates, such as bacterial mass generated during biological treatment, are separated from the liquid phase by sedimentation, filtration, or centrifugation. Soluble and colloidal products of bacterial metabolism, however, tend to remain in the bulk liquid.

Bacterial growth can be supported by a wide variety of organic compounds that influence biological treatment in a number of ways. Many organic materials are used by organisms, while other types of organics, such as synthetic organic chemicals (*xenobiotics*), resist biological degradation (*persistence*) or may be toxic to organisms and treatment systems. Metabolism can be inhibited or stopped completely in the presence of even low concentrations of certain toxic compounds in wastewater. However, organics that are toxic or inhibitory at high concentrations can be used at lower concentrations. Frequently, organisms are capable of adapting or acclimating to higher concentrations of toxic compounds if sufficient time and adequate environmental conditions are provided.

There are two major objectives to this chapter. The first is to review the factors and processes influencing the fate and behavior of organic compounds in biological wastewater treatment systems. The second is to examine why some organic compounds are readily removed by biological treatment systems, while the response to other organics may vary from partial or no degradation to partial or total inhibition.

The characteristics of organic compounds are examined, and techniques for measuring the concentration of organic compounds and identifying specific compounds are reviewed. Properties and processes that influence the fate of organic compounds are described. Finally, general aspects of the fate and removal of organics in wastewater treatment processes are discussed.

SOURCES OF ORGANIC COMPOUNDS IN WASTEWATER

Organic compounds in wastewater originate from a variety of sources. Domestic wastewater includes naturally occurring organic matter of vegetable, animal, and human origin, and *anthropogenic* (human-made) compounds such as soaps, synthetic detergents, and household chemicals. Commercial or industrial effluents include organic waste materials, products, and organic chemicals discharged from industrial process operations. Depending on the wastewater source, organic compounds may be present as major components of wastestreams or as minor components at trace levels.

Additional sources of organic compounds in wastewater are contaminants present in municipal water supplies. For example, in some areas, water supplies contain low levels of herbicides, insecticides, pesticides, or industrial wastes from runoff or groundwater infiltration.

Toxic organic chemicals can cause inhibition of biological wastewater treatment, even if readily usable organics are available and toxic compounds are present at only trace levels. The composition of wastewater is not constant and contains organic compounds in dissolved, colloidal, and particulate forms. However, some toxic compounds can become associated with colloidal or particulate matter through sorption (absorption or adsorption) processes, reducing their bioavailability and toxicity to biological activities. Industrial wastes can be discharged continuously or intermittently. Either way, *shock loads* (high loads of organic compounds, particularly toxic organic compounds) can cause process upsets and instability.

TOXIC ORGANIC CHEMICALS. Contaminants and pollutants are *toxic* if they have the potential to inhibit biological treatment processes, adversely affect the health and welfare of humans, or cause environmental damage to fish, wildlife, or any other sensitive component of an ecosystem. Effects of toxic chemicals depend on the specific chemical and the route and duration of exposure. Some toxic organic chemicals interfere with normal metabolism and cause permanent physiological damage, such as neurotoxins on brain tissue. Other types of chemicals cause kidney, liver, or lung damage.

Chemicals called *mutagens* cause alterations of the genetic code and interfere with normal development. Other organic chemicals can damage the reproductive cycle and cause permanent damage to developing infants. The damage can be evident at birth as birth defects or can express itself through some type of system malfunction later in life. Such chemicals that cause reproductive effects in offspring but not in parents are called *teratogens*. Carcinogens are toxic chemicals that cause cancer from long- or short-term exposure.

Determining which chemicals are toxic and to what extent are difficult tasks. Extensive tests are required to determine the concentration level at which a chemical can pose serious risks. Because there are many potential manifestations of toxicity, it is difficult to identify all of the toxic properties of a chemical and to quantify risk from each possible route of exposure. Some chemicals are only toxic when exposure occurs in the presence of other chemicals (*mutualistic*) or after they have been metabolized (*bioactivation*).

In many cases, a chemical is widely used before its toxic characteristics are discovered. Cancer development can take up to 20 years, so the cancer risk from exposure to a chemical may not be known until long after it has been used. Also, it can be difficult to prove a specific chemical as responsible for a disease.

For regulatory purposes, toxicity is assessed on the basis of chronic (long-term) or acute (short-term) exposure. Toxicity is rated through a series of laboratory, toxicological, and epidemiological tests. Maximum exposure limits are determined based on levels of acceptable risk. Wastewater treatment goals are derived from chemical toxicity data and are based on the quality of water received and on the potential risk of health or environmental damage.

Toxic chemicals find their way to wastewater treatment facilities through industrial wastewaters, domestic and commercial use of chemicals, and surface runoff. Although toxic contaminants can inhibit biological wastewater treatment systems, often they are metabolized when present at low concentrations or when a sufficiently long acclimation period for the microbial population has been provided. The toxicity of an individual synthetic organic compound is the result of its chemical structure and concentration in a biological reactor. Large chemical concentrations may be necessary to produce an inhibitory or toxic effect on the treatment process. Inhibitory concentrations of some toxic organic compounds range from 20 µg/L to 500 mg/L (Anthony and Breinhurst, 1981). Chemicals can act independently or their combined toxic effects can be additive, synergistic, or *antagonistic* (one chemical masks the toxic effect of another).

CLASSIFICATION OF ORGANIC COMPOUNDS

Organic compounds consist of carbon *covalently bonded* (shared electron pairs) to one or more elements. Carbon, hydrogen, and oxygen are the major elements composing organic compounds, while nitrogen, phosphorus, and sulfur are minor elements. Organic compounds also can contain *halogens* (nonmetallic elements), metals, and other elements. The major organic constituents of municipal wastewater are proteins, carbohydrates, and lipids. *Humic* (soil) compounds, detergents, solvents, and agricultural and industrial chemicals also can be present in wastewater. Properties that differ among the chemical components of wastewater include molecular size, solubility, reactivity, and toxicity.

The chemical structure of organic compounds influences their behavior in wastewater. Certain properties of organic compounds influence their ability to participate in the biological reactions that occur in wastewater treatment. Functional groups of organic compounds are important in determining the fate of a chemical in a biological treatment system, and those commonly found in organic compounds are shown in Table 8.1. Functional groups are attached to one of the carbon atoms in organic compounds and are locations of either high or low electron density. They provide sites for reactions that can occur between functional groups of different organic compounds. These reac-

Table 8.1   Organic functional groups.

Functional group		Type of molecule	Example	
Name	**Formula**		**Name**	**Formula**
Methyl	(methyl structure)	Hydrocarbon and others	Methane	(methane structure)
Hydroxyl	—OH	Alcohols	Propanol	(propanol structure)
Carbonyl	(carbonyl structure)	Aldehydes and ketones	Glyceraldehyde	(glyceraldehyde structure)
			Dihydroxy acetone	(dihydroxy acetone structure)
Carboxyl	(carboxyl structure)	Acids	Acetic acid	(acetic acid structure)
Amino	(amino structure)	Amino acids and amines	Glycine	(glycine structure)
Sulfhydryl	—S—H	Certain amino acids, mercaptans	Cysteine	(cysteine structure)

tions occur because of the attractive forces between regions of high electron density in one molecule and regions of low electron density in another.

One of the most common systems of classifying organic compounds is based on their molecular structure. Brief descriptions of the major characteristics of organic compounds that can be present in wastewater are listed in Table 8.2. Organic compounds differ on the basis of molecular weight, *polarity* (charge distribution), and volatility. The common structures of aliphatics, alcohols, ethers, aldehydes, ketones, esters, carboxylic acids, and amines are of low molecular mass compared to larger molecules such as proteins, polymeric carbohydrates, and lipids. The low molecular mass compounds often are subunits or metabolic products derived from biological metabolism of larger molecular mass compounds.

Table 8.2   Characteristics of organic compounds present in wastewater.

Type of compound	Characteristics	Chemical structure	Example
Aliphatic hydrocarbons	Contain carbon and hydrogen		
Saturated	Single bond between carbon atoms	$C_nH_{(2n+2)}$	Ethane, propane
Unsaturated			
Ethene series	Double bond between carbon atoms	C_nH_{2n}	Ethylene
Ethyne series	Triple bond between adjacent carbon atoms	C_nH_{2n-2}	Acetylene
Alcohols	Contain a hydroxyl group	R—OH	Ethyl alcohol
Ethers	Contain oxygen between two carbon atoms	R—O—R	Ethyl ether
Aldehydes	Contain a double bond between a carbon atom and oxygen	R—C—O $\vert$ H	Solvents, Formaldehyde
Ketones	Contain at least two organic groups	R—C—O $\vert$ R	Acetone
Organic acids	Contain a carboxylic acid	OH $\vert$ R—C—O	Acetate
Esters	Contain a carboxylic acid and at least two organic groups	O $\vert$ R—C—O—R	Ethyl formate

Type of compound	Characteristics	Chemical structure	Example
Amines	Contain an organic group and an amine group	$R_nNH_{(3-n)}$	Methyl amine
Cyclic compounds	Contain at least one ring structure		
Aromatic compounds	Contain a benzene ring		Benzene
Phenols	Contain a benzene ring and a hydroxyl group		Phenol
Methyl derivatives	Contain a benzene ring and a methyl group		Toluene, xylene
Halogen derivatives	Contain a benzene ring and a halogen		Chlorobenzenes
Heterocyclic compounds	Contain a ring structure containing carbon and at least one other element in the ring		Naphthalene
Proteins	Organic acids with an amine group attached to a chain containing an acid group, 16% nitrogen		Enzymes, animal and plant proteins
Carbohydrates			
Monosaccharides	Simple sugars		Glucose, triose, pentose, fructose, hexose, galactose
Disaccharides	Can be hydrolyzed to simple sugars	$C_{12}H_{22}O_{11}$	Sucrose, maltose, lactose
Polysaccharides		$(C_6H_{10}O_5)_n$	Starch, cellulose
Lipids			Fatty acids, soaps, detergents

*T*OXIC ORGANIC COMPOUNDS

The chemical structure of a compound influences its toxicity and behavior in wastewater treatment. Among the more common groups of toxic compounds found in wastewater are aliphated hydrocarbons, halogenated aliphatic hydrocarbons, aliphatic alcohols, aromatic hydrocarbons, and halogenated aromatic hydrocarbons.

ALIPHATIC HYDROCARBONS. Aliphatic hydrocarbons are straight chain hydrocarbons. If they contain fewer than four carbon atoms, these compounds exist as gases such as methane (CH_4), ethane (C_2H_6), propane (C_3H_8) and butane (C_4H_{10}). Higher molecular mass hydrocarbons exist as liquids and often are found in wastewater. Aliphatic hydrocarbons with five to eight carbon atoms often are used as solvents and include pentane (C_5H_{12}), hexane (C_6H_{14}), heptane (C_7H_{16}), and octane (C_8H_20). Gasoline and kerosene contain aliphatic hydrocarbons and others such as branched, *unsaturated* (carbon atoms have doubled bonds), and *aromatic* (ring) hydrocarbons.

HALOGENATED ALIPHATIC HYDROCARBONS. Halogenated aliphatic hydrocarbons contain nonmetallic elements and have excellent solvent power and low flammability. These compounds tend to have toxic effects, including depression of the central nervous system and the potential to cause kidney damage. Examples of halogenated aliphatic hydrocarbons include methylene chloride, chloroform, carbon tetrachloride, methyl chloroform (1,1,1 trichloromethane), trichloroethylene, and tetrachloroethylene. These chemicals are widely used industrial solvents and have been identified in wastewater effluent.

ALIPHATIC ALCOHOLS. Aliphatic alcohols are straight or branched chain alcohols used as industrial solvents. These compounds include methanol, ethanol, and *n*-butanol.

AROMATIC HYDROCARBONS. Aromatic (ring) hydrocarbons have widespread use as solvents in the chemical and drug industries. These solvents tend to be volatile and toxic. Aromatic hydrocarbon solvents include benzene and toluene.

HALOGENATED AROMATIC HYDROCARBONS. Halogenated aromatic hydrocarbons are stable compounds with low flammability and high toxicity. Examples of these compounds are polychlorinated biphenyls (PCBs) used as insulating materials in electrical capacitors and transformers, chlorophenols such as pentachlorophenol and hexachlorophene, and dioxin or 2,3,7,8- tetrachlorodibenzo-*p*-dioxin (TCDD).

Toxic compounds frequently detected in wastewater effluent are listed in Table 8.3. Chlorinated aliphatic and chlorinated aromatic solvents are the groups most frequently detected.

SIGNIFICANCE IN WASTEWATER TREATMENT

The chemical structure of an organic compound determines its physical and chemical properties, which influence the fate and conversion of organic

Table 8.3 Toxic compounds most frequently reported in municipal effluent (Pellizari and Little, 1980).

Reported frequency ranking	Compound	Reported frequency ranking	Compound
1	Tetrachloroethene	26	1,1-Dichloroethene
2	Dichloromethane	27	o-Ethyltoluene
3	Trichloroethene	28	Benzoic acid
4	2-(2-Butoxyethoxy)ethanol	29	2-N-Butoxyethanol
5	Benzene	30	Dimethyldisulfide
6	Toluene	31	Diethyl-o-phthalate
7	Chloroform	32	Lauric acid (N-dodecanoic acid)
8	Ethylbenzene	33	1,8-Dimethyldisulfide
9	1,1,1-Trichloroethane	34	2-Propanone (acetone)
10	Phenol	35	Tetradecanoic acid
11	p-Cresol	36	Decanoic acid
12	Caffeine	37	Methylisobutylketone
13	m-Cresol	38	2,7-Dimethylnaphthalene
14	Cycloheptatriene	39	n-Pentadecane
15	Octadecane	40	Dibutylphthalate
16	Phenylacetic acid	41	1-Hydroxy-2-phenylbenzene
17	Dioctylphthalate	42	1,2,4-Trimethylbenzene
18	1,4-Dimethylbenzene (p-xylene)	43	Indole
19	1-Methylnaphthalene	44	n-Hexane
20	m-Xylene	45	n-Eicosane
21	Hexadecane (practical)	46	Dioctylphthalate
22	2-Methylnaphthalene	47	1,3-Dimethylnaphthalene
23	o-Cresol	48	p-Ethyltoluene
24	a-Terpineol	49	2,4-Dimethylphenol
25	Naphthalene	50	1,3,5-Trimethylbenzene

chemicals in biological treatment processes. Properties such as solubility, vapor pressure, diffusivity, n-octanol and water partition coefficient, and sorption partition coefficient influence the transfer, partitioning, and exchange of organics among liquid, solid, and gaseous phases. Chemical properties also influence physical removal of organics by *volatilization* (evaporation) or *sorption* (adherence) on waste sludges. The ability of an organic compound to participate in biological or nonbiological reactions such as *oxidation* (loss of electrons), *reduction* (gain of electrons), *hydrolysis* (water cleavage), *photolysis* (light cleavage), and *polymerization* (forming more complex molecules) is controlled by chemical structure and properties.

The biodegradability of organic compounds is influenced by their physical and chemical properties. Many organic compounds, including alcohols, carboxylic acids, carbohydrates, and amino acids, are readily biodegradable under both aerobic and anaerobic conditions. Chlorinated aliphatic compounds (trichloroethylene) tend to be more degradable anaerobically, while chlorinated aromatics tend to be more degradable aerobically. From a biological process point of view, the characteristics of organic compounds affect the significance of biodegradation as a removal mechanism.

Bacterially mediated reactions, the most significant removal mechanism for readily degradable organics in wastewater treatment, often are the most significant removal mechanism for recalcitrant compounds (Barton, 1987; Namkung and Rittmann, 1986; and Weber *et al.*, 1987). Toxicity of wastewater components and inhibition are other critical characteristics influencing removal of organic contaminants. If toxic constituents that inhibit biological treatment are present, degradation of nontoxic organics can be impaired and the effectiveness of treatment can be reduced dramatically.

MOLECULAR SIZE RANGES OF ORGANIC COMPOUNDS

Organic contaminants in wastewater include dissolved organic compounds, colloidal particles smaller than 1 µm, supracolloidal particles larger than 1 µm, and settleable particles. Treatment plant influents typically contain organic matter ranging in size from less than 0.001 to more than 100 µm. Material comprising the biochemical oxygen demand (BOD) and volatile suspended solids (VSS) of settled municipal wastewater typically is smaller than 50 µm. Approximate size ranges of typical organic constituents characteristic of settled municipal wastewater are presented in Figure 8.1. The standard suspended solids (SS) test involves filtration through a glass fiber filter of 1.2-µm nominal pore size. This material includes protozoa, algae, bacterial flocs and single cells, waste products, and other debris. Particulate matter smaller than 1.2 µm is typically not detected by the standard SS test.

Organic particles smaller than 0.1 µm are typically cell fragments, viruses, macromolecules, and debris. The major groups of macromolecules in wastewater are polysaccharides, proteins, lipids, and nucleic acids. Because of difficulties associated with the measurement of molecular size, macromolecules typically are defined by apparent molecular mass. The apparent molecular mass is measured in atomic units (1 amu = 1.66×10^{-24} g). The apparent molecular mass range of macromolecules found in wastewater is between 10^3 and 10^6 atomic mass units. As shown in Figure 8.1, a macromolecule with a molecular mass of approximately 10^6 amu corresponds to a particle size of approximately 0.01 µm. Compounds in the range from 10^3 to 10^6 amu in-

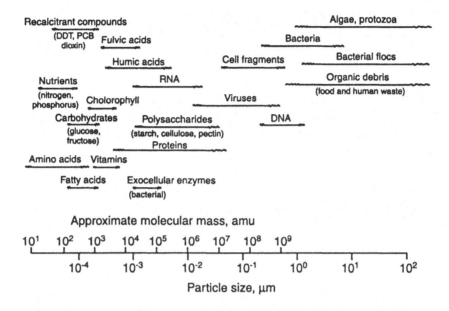

Figure 8.1   Typical organic constituents in settled municipal wastewater (Levine *et al.*, 1985)

clude humic and fulvic acids that may be present in the water supply and in sewer inflows and infiltration. Wastewater compounds smaller than 10^3 amu include carbohydrates, amino acids, fatty acids, vitamins, and chlorophyll.

QUANTIFICATION OF ORGANIC COMPOUNDS

There are a number of ways in which organic compounds in wastewater can be measured, based on different properties of the compounds. Organics can be measured collectively using nonspecific analyses to determine the overall content in the wastewater. Alternatively, individual compounds can be identified and measured using specific analyses. The use of nonspecific analyses is advantageous because an overall measure of the level of organic compounds present is provided, though identification of the individual compounds is not possible.

Specific analyses are useful when information on the identity and concentration of individual organic compounds is needed. Specific analyses are more

---

sensitive than nonspecific analyses and can be used to measure concentrations in the μg/L (parts per billion) range. Many specific analyses for organic compounds are difficult and expensive to conduct, thus are used primarily for compliance monitoring and research purposes. Nonspecific analyses, which are used in treatment plant operation and control, are less sensitive and useful in measuring organic content in the mg/L (parts per million) range.

NONSPECIFIC MEASURES. A summary of nonspecific tests used to conduct nonspecific analyses of organic matter in wastewater is given in Table 8.4. Each test measures a specific property of organic compounds, but many lack the sensitivity or specificity to detect the presence of toxic organics, largely because the toxic chemical may be present at trace levels, contributing insignificantly to the property being measured. The types of compounds measured by each test and the limitations of each procedure are important considerations.

Biochemical Oxygen Demand. Biochemical oxygen demand (BOD) is a parameter widely used in wastewater treatment. It is used as a design parameter, for process control and efficiency monitoring, and to meet permit requirements.

The BOD is the quantity of oxygen required by organisms to oxidize organic carbon to carbon dioxide and organic nitrogen to nitrate. Biochemical oxygen demand is subdivided into carbonaceous biochemical oxygen demand (CBOD) and nitrogenous biochemical oxygen demand (NBOD).

The BOD of a sample is determined by incubating the wastewater in the presence of large quantities of oxygen and bacteria at a constant temperature for a fixed time period. The oxygen consumed during the reaction period is the BOD of the sample. If nitrification inhibitors are used, then the amount of oxygen consumed is the $CBOD_t$, where t is the number of days the sample was incubated.

The rate of microbial oxidation of organic matter in the BOD test depends on the type of organic matter present in the wastewater, the type of organisms used in the test, the presence of nutrients, the pH and temperature, and the presence of toxic constituents. The maximum amount of oxidation that can occur is called the ultimate BOD or BOD_u.

There are problems with using CBOD as a measure of organic content of a wastewater sample. Oxidation rates tend to be slow, and a lag phase often is observed with this test. To save time, the CBOD test has been standardized and, in most cases, is measured after 5 days of incubation ($CBOD_5$). A plot of the CBOD reaction as a function of time is shown in Figure 8.2 for two samples with the same BOD_u but a markedly different $CBOD_5$. If analyzed too early during the incubation period, the CBOD measurement may be misleading.

Another problem with using $CBOD_5$ as a measure of organic content can arise if toxic constituents are present in the wastewater. Toxic compounds tend to inhibit biological activity and suppress the rate of oxidation of nontoxic organic matter present in the wastewater. The $CBOD_5$ for such a sample

Table 8.4 Summary of nonspecific tests used to quantify organic matter in wastewater.

Test	Description
Oxygen demand	
Biochemical oxygen demand (BOD)	
Carbonaceous ($CBOD_t$)	The quantity of oxygen required by organisms to oxidize organic carbon at a constant temperature and for a time period, t.
Nitrogenous ($NBOD_t$)	The quantity of oxygen required by nitrifying bacteria to oxidize organic nitrogen at a constant temperature and for a time period, t.
Ultimate (BOD_u)	The maximum amount of oxygen required to biochemically oxidize organic matter in wastewater at a constant temperature and for an infinite time period.
$BOD_u = CBOD_u + NBOD_u$	
Chemical oxygen demand (COD)	The equivalent quantity of oxygen necessary to chemically oxidize organic carbon to carbon dioxide.
Theoretical oxygen demand (ThOD)	The quantity of oxygen required to completely oxidize all organic carbon and organic nitrogen in a sample.
Carbonaceous oxygen demand (CThOD)	The quantity of oxygen required to completely oxidize all organic carbon in a sample to carbon dioxide.
Nitrogenous oxygen demand (NThOD)	The quantity of oxygen required to completely oxidize all organic nitrogen in a sample to nitrate.
$CThOD + NThOD = ThOD$	
Total organic carbon	The quantity or organic carbon that can be oxidized in the presence of a strong oxidant or by combustion.
Total organic halogen	The quantity of organic halogen that is present in a wastewater sample.
Ultraviolet absorption	The quantity of organic carbon in a wastewater sample that absorbs light in the ultraviolet range.
Volatile solids	The quantity of organic matter that is combusted at 550°C (1 020°F).

may be artificially low due to toxic interference with the biological conversion process; thus, the actual content of degradable organics is not reflected as illustrated in Figure 8.3.

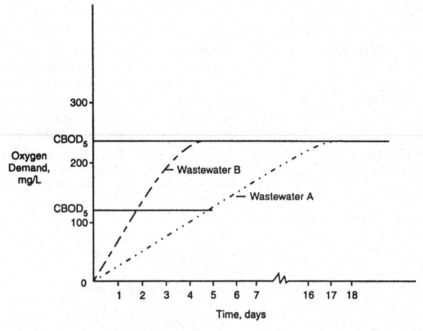

Figure 8.2 **Comparison of CBOD₅ for two wastewaters with the same CBOD_u.**

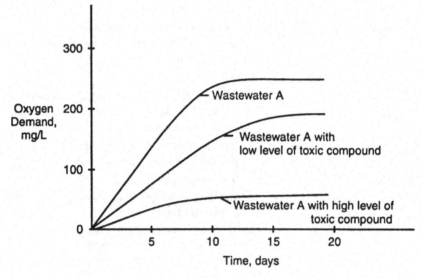

Figure 8.3 **The influence of toxic compounds on CBOD measurement of wastewater. The three lines represent samples of a municipal wastewater with a COD value of 375 mg/L.**

Chemical Oxygen Demand. The chemical oxygen demand (COD) is used as a measure of the quantity of organic matter that can be oxidized by a strong chemical oxidant in the presence of sulfuric acid. The standard procedure for this test uses dichromate (Cr_2O_7) to oxidize organic matter by the following general equation:

$$\text{Organic matter} + Cr_2O_7^{--} + H^+ \rightarrow Cr^{3+} + CO_2 + H_2O \qquad (8.1)$$

In this chemical test, it is assumed that the quantity of oxygen needed for complete oxidation of organic carbon is proportional to the quantity of dichromate consumed in the oxidation reaction. Actual chemical oxidation of most organic compounds in wastewater is at least 95% of the theoretical value. For organic compounds that are biologically degradable in the presence of oxygen, the $CBOD_u$ is nearly equivalent to the COD. Thus, the ratio of $CBOD_u$ to COD is an indication of the degree of aerobic biodegradability.

Not all organic compounds are oxidized in the COD test. Volatile organic compounds are oxidized only to the extent that they do not volatilize (evaporate) and exit from the test vessel. Pyridine compounds resist oxidation. Ammonia (NH_3) present either in the aqueous sample or released from nitrogen containing organic matter is oxidized only when free chloride ions are present. Aqueous samples from methanogenic systems contain dissolved gases such as methane (CH_4) and hydrogen sulfide (H_2S), which also have a chemical oxygen demand. The degree to which dissolved gases contribute to COD depends on the sample chemistry and must be verified by using control solutions and independent measures. It should be noted that the results can be difficult to interpret. Trace levels of organic compounds are not quantified by the COD test. In general, oxidized organic matter at concentrations greater than 5 mg/L as oxygen can be measured reproducibly.

Theoretical Oxygen Demand. The theoretical oxygen demand (ThOD) is the quantity of oxygen needed to oxidize all organic carbon in a sample to carbon dioxide and all organic nitrogen to nitrate. The two major components of the ThOD are the carbonaceous oxygen demand (CThOD) and the nitrogenous oxygen demand (NThOD). These values can be determined if the exact chemical formulas for the organic compounds in wastewater are known.

Total Organic Carbon. The total organic carbon (TOC) of a sample is a measure of the quantity of organic carbon that can be chemically or thermally oxidized to carbon dioxide. Determining the TOC of a wastewater sample is a fairly rapid test and is a reproducible measure of the quantity of organic carbon. Thus, TOC can be useful for process monitoring in a wastewater treatment system.

For samples containing organic compounds with known chemical compositions, the organic carbon content can be calculated directly. However, some types of compounds are not oxidized completely in the TOC procedure. Another disadvantage is that, like the COD test, the TOC test cannot detect trace levels of organic compounds because of its limited sensitivity.

Total Organic Halogen. The total organic halogen (TOX) test is a measure of the concentration of halogen species (Cl^-, Br^-, F^-) associated with organic compounds in an aqueous sample. This test is conducted using an instrument containing two columns of activated carbon in series. After the organic compounds are adsorbed (adhere) to the carbon column and the sample is chemically decomposed by heat (*pyrolyzed*), the quantity of halogen associated with the sample is detected. This test is more sensitive than the TOC test but can only detect halogenated compounds. As there are many organic compounds that are toxic but not halogenated, the TOX test is not a comprehensive measure of toxicity.

Ultraviolet Absorption. Another nonspecific analysis is the use of ultraviolet (UV) absorption. This test is simple to conduct, and the quantity of organic matter that is UV absorbable can be correlated to the TOC of a sample from a given location and level of treatment. Therefore, measuring UV absorbance can provide a rapid measure of the organic content of a sample, although it is difficult to compare UV absorbance data from different locations or from different stages of treatment. The optimum absorbance depends on the chemical composition of the wastewater.

Volatile Solids. The volatile solids test is a measure of the quantity of solids combusted from a wastewater sample at 550°C (1 020°F). If it is assumed that only organic carbon would be converted to CO_2 at 550°C, then the volatile solids content can be related to the content of organic carbon. If the wastewater contains a high concentration of dissolved carbonates, these may also be measured as volatile solids.

SPECIFIC ANALYSES. Individual organic compounds can be analyzed by a variety of techniques. The preferred technique for detection and quantification of organics from wastewater is determined by the volatility, *polarity* (charge distribution) and molecular weight of the compound. The measurement procedures include the following steps:

- Isolation or extraction of organic compounds from wastewater or sludge;
- Concentration of the organic mixture to measurable levels;
- Separation of the contaminant of interest from other contaminants or interferents;
- Detection and identification of the compounds; and
- Quantification of the compound.

Chromatography. *Chromatography*, the most widely used method for quantification of organic compounds, is a process of separation and quantification of organic molecules in a solute mixture such as wastewater. Separation is achieved by distributing the solute mixture between a mobile phase and a stationary phase. Solutes preferentially distributed in the mobile phase will more readily move with the mobile phase, while movement of solutes preferentially distributed in the stationary phase will be retarded.

The size, polarity, and volatility of a compound influence its partitioning behavior. Chromatographic separation is achieved through selection of stationary and mobile phases that best exploit the different molecular forces existing between solute molecules and each of these phases. Forces that control solute retention can be ionic forces resulting from the electrical charge of the solute, polar forces that arise from unequal charge distribution among the solute particles, or electrical dispersive forces among molecules.

The most effective chromatographic method used for separations and quantitative analysis of organic compounds is elution chromatography. In this mode of chromatography, component mixtures of organics in wastewater separate into discrete bands representing each component or solute. The bands are *eluted* (passed through) at different rates, forming chromatographic peaks. The integrated area of each chromatographic peak is compared to that of a standard solution of known concentration.

Chromatographic analysis of organic compounds is essentially a comparison technique rather than a definitive identification technique. The type of chromatographic column to be used depends on the characteristics of the types of compounds to be analyzed and operating conditions such as temperature and pH can be optimized to improve separation of compounds. The sample preparation can also influence the effectiveness of a particular column to separate specific compounds. However, procedures for most organic compounds in wastewater have been standardized and, within a given system and under specific operating conditions, retention time in the column is the only factor used to separate different compounds. Therefore, it is possible to have compounds other than the contaminant of interest to elute at the same retention time. For example, it has been reported that many PCBs cause positive interferences with the gas chromatography (GC) analysis of chlorinated hydrocarbons. Positive identification thus requires confirmatory analysis with an additional chemical or instrumental method.

Gas Chromatography. *Gas chromatography* is a separation technique that relies on the differential partitioning of organics to a stationary liquid phase and a mobile gas phase. The mobile gas phase is used to sweep a mixture of volatile organic compounds along a chromatographic column containing the stationary phase. Traditional GC columns are long, small-diameter stainless steel or glass columns packed with a solid support such as silica gel, onto which is coated the stationary phase. An immobile, waxy liquid forms the stationary phase. Different compounds partition between the moving stream of gas and the liquid phase at different rates and the mixture is resolved into narrow bands of individual components. These bands then pass through a detector. Identification and quantification are based on comparison of the detector response to the response of standard compounds of known concentrations.

Chromatographic System. The gas chromatographic system consists of a pneumatic controller (a device containing compressed gas) that provides carrier gas flow for the column and detector gases, an injection system consisting of either a septum port or a sample valve, the column, the column oven that allows for operation at either a constant temperature (*isothermal*) or temperature programming, and the detectors and associated electronics.

Gas chromatography can be performed by direct injection of gas or liquid samples, though sample preparation often is required. Samples may need to be concentrated or the solutes may need to be extracted from a liquid form into a solvent. The extraction, concentration, and cleanup steps before instrumental analysis can be time consuming.

Much of the work is labor intensive, requiring specialized glassware, high-purity solvents, a well-equipped laboratory, and trained staff. Testing laboratories typically require more than 2 weeks for sample turnaround time, and the monitoring of organics for process control is not easily accomplished under these conditions.

Detection Devices. Many types of detection devices are available for the identification of separated organic compounds, all of which are based on measuring a property that differs between the pure carrier gas stream and the eluant and gas stream mixture. Detectors can be classified in a number of ways. The detector response can be universal and respond to any compound or the response can be specific and measure some particular property, element, or structure of a solute molecule. Detectors are also classified by their ability to detect compounds according to either compound mass or concentration.

Combined Gas Chromatography and Mass Spectrometry. The mass spectrometer detector (MSD) is an independent analytical instrument more complex than the GC itself. Through electron impact or chemical ionization, the MSD separates, quantifies, and identifies ions from GC eluates by generating a full mass spectrum from minute quantities of a compound. Molecules are ionized and measured according to their mass-to-charge ratio. Mass-to-charge ratios of ionic products are unique for each compound and can be used to identify organic compounds by comparing the derived mass spectrum to computerized library values. Both GC and MS often are used for confirmation analyses of the analytical results from other GC columns or detectors and system configurations.

U.S. EPA lists methods for GC analysis for a variety of toxic organics. Each method is a standardized procedure for identifying and quantifying a specific group of organic compounds. To properly conduct each method, specialized equipment is required. Specific requirements for column configuration, detector to be used, and sample preparation are detailed in each method. Recent improvements in gas chromatography, such as the use of capillary columns, have made quantification of trace organics at the μg/L level feasible in many cases.

High-Performance Liquid Chromatography. A liquid chromatograph (LC) consists of a solvent programmer, a high-pressure pump, and a mobile-phase supply unit that supplies from one to four solvents. Once the LC sample is placed on the column, the separated organic compounds are identified with a detector.

Samples to be analyzed typically require sample preparation, which can include simple dilution with an appropriate solvent, *derivatization* (modifying the chemical to allow detection and quantification), or other preprocessing of the sample. Many liquid samples require one or more treatments, such as solvent extraction, concentration, filtration, or derivitization, depending on the particular analysis.

The most widely used detectors for quantitative high-performance liquid chromatography (HPLC) are the ultraviolet absorption detector (UV), refractive index detector (RI), fluorescence detector, and the electrochemical detector. In the UV detector, ultraviolet light is directed through the column *eluant* (sample) and the intensity of the light absorbed by the solutes is monitored. The light source can be either a fixed wavelength or a broad band of wavelengths, and only compounds that absorb UV radiation are detected.

The RI detector monitors the change in angle as a light beam is refracted upon passing from one medium to a second, adjacent medium—the LC eluant stream. Solutes in the eluant cause the refractive index of the eluant stream to change, and the direction of the light beam is changed. The RI detector is practically a universal detector for LC, as almost any solute will alter the refractive index of the liquid mobile phase. However, it is sensitive to temperature fluctuations. The fluorescence detector monitors light emitted by solute molecules in response to absorbance of UV light, while the electrochemical detector reflects the oxidative or reductive tendency of eluting solute molecules. Both the fluorescence and the electrochemical detectors are sensitive but also specific and are not as widely used as the UV detector.

High-performance liquid chromatography is a powerful analytical technique applicable to a wide variety of organic compounds. Some organic chemicals of environmental concern that can be quantified by HPLC include chlorinated pesticides, organophosphate pesticides, triazine herbicides, polynuclear aromatic hydrocarbons, benzidines, carbofuran, aldicarb and sulfone and sulfoxide derivatives, aliphatic benzenes, phenols, phthalate esters, aromatic nitro compounds, carbohydrates, and proteins.

Thin-Layer Plate Chromatography. Thin-layer plate chromatography
(TLC) is performed using a flat, rigid plate coated with a thin layer of the stationary phase sorbent, with a trough along one side of the glass plate containing the mobile phase. The flat plate is prepared with a homogeneous layer of
small particles of narrow particle size distribution. Stationary phases typically
consist of various forms of silica, alumina, or cellulose. The mobile phase migrates from the trough across the flat plate stationary phase. Samples containing the organics for analysis are applied to the stationary phase as spots or
bands using mechanical applicators or streaking devices.

The samples are placed along the edge of the flat plate where the mobile
phase trough is located. As the mobile phase migrates across the samples and
across the plate, organic compounds having greater affinity for the mobile
phase will move farther across the plate, and chromatographic separation of
sample solutes is achieved.

Automatic scanning densitometry is a method used to determine spot location and to quantify the concentration of the separated compound. Polycyclic
aromatic hydrocarbons are one group of organic compounds amenable to
TLC analysis.

ORGANIC COMPOUNDS IN BIOLOGICAL SYSTEMS

Predicting the fate of organic compounds in biological treatment systems is not
without difficulty. The competing mechanisms of removal typically consist of
biological degradation, volatilization to the gas phase, and adherence to solids.
Both physical and chemical properties of organic compounds influence their
susceptibility to removal by these mechanisms. Some of the more influential
of these properties include solubility, diffusivity, vapor pressure, octanol or
water partition coefficient, and adsorption coefficients. These properties influence the propensity of organic compounds to transfer to other phases, or to be
transformed by chemical, photochemical, or biological reactions.

SOLUBILITY. *Solubility* is the maximum concentration of a chemical that
will dissolve in pure water at a specified temperature. Above this solubility
limit, two separate liquid phases will exist if the organic compound is a liquid
at the system temperature. The solubilities of organic chemicals are expressed
in micrograms per liter (μg/L), and range from approximately 1 to 10^5 μg/L at
ambient temperatures, though some can be much higher. The solubility of a
compound is determined by adding the chemical to pure water and allowing
it to equilibrate at a constant temperature. For *hydrophobic* compounds (those
that do not hydrate easily), this step can take several days. The solution is fil-

tered or centrifuged to remove undissolved material and the concentration of dissolved organic compound is measured.

Solubility is influenced by hydrogen ion activity (pH), temperature, and ionic strength. Most compounds become more soluble with increasing temperature or ionic strength, whereas the influence of pH depends on that particular compound. The presence of high molecular mass compounds such as humic and fulvic acids tends to increase the solubility of many organic chemicals.

Water solubility influences the fate of chemicals in wastewater treatment. Highly soluble chemicals are more mobile, thus more likely to be found in the water phase. Such chemicals are less likely to sorb to organic particles, and are more available for transformation by organisms. They also are less likely to be removed by separation of biological solids onto which they have sorbed. Water solubility influences degradation reactions such as oxidation, photolysis, and hydrolysis.

HENRY'S LAW CONSTANT. The Henry's law constant is an equilibrium ratio of the concentration of a gas (kilopascals [atmospheres] per mole) that will dissolve in a given volume of liquid (litres [gallons]) at a constant temperature (degrees Celsius [Fahrenheit]), to the partial pressure (kilopascals [atmospheres]) that the gas exerts above the liquid (Stumm and Morgan, 1981). For a low solubility gas such as methane (CH_4), which has a Henry's law constant of 665 atm-litre/mole at 25°C (75°F), the solution concentration in equilibrium with one atmosphere of CH_4 would be:

$$CH_4 = 1 \text{ atm}/665 \text{ atm-litre/mole} = 0.001\ 5 \text{ moles/litre} \qquad (8.2)$$

Henry's law indicates how far a gas or liquid system is from equilibrium. A compound with a high Henry's law constant will be more likely to volatilize into the gas phase. The Henry's law constant thus influences the rate of gas transfer and the removal of gases or volatile compounds from liquids. Measurement of Henry's law constant is accomplished through use of aeration devices that typically involve transfer of gases near atmospheric pressure to either air bubbles rising through liquid or to thin films of liquid flowing over surfaces exposed to air.

DIFFUSIVITY. *Diffusion* is a measure of the transport of molecules by intermolecular collisions in nonturbulent liquid or gas media. The driving forces for diffusional processes are concentration gradients both in vapor, liquid, and solid phases, and across media interfaces. The propensity for diffusion of an organic molecule is represented by the diffusion coefficient, or diffusivity.

Diffusion can occur in the vapor phase, the aqueous phase, across the vapor or liquid interface, or across the solid or water interface. The rate of diffusion depends on the diffusion coefficient for the diffusing solute in water, the system temperature, and the concentration gradient of molecules that drives the process. Diffusion coefficients increase with temperature.

Diffusion in a gaseous phase is the most rapid, while liquid- and solid-phase diffusion are slower. Diffusion coefficients influence mass transfer of compounds between phases, which affects the rate at which organic compounds are volatilized and the availability of organic solutes to organisms.

VAPOR PRESSURE AND VOLATILITY. The vapor pressure of a liquid is measured as the pressure exerted by a saturated vapor in equilibrium with a liquid at a given temperature. The vapor pressure of water and organic liquids increases with temperature. Vapor pressure is a chemical-specific property that allows estimation of the rate of evaporation and can influence the volatility and persistence of sorbed organic compounds.

Volatilization is the process by which a compound evaporates to the vapor phase from the solution (liquid) phase. The factors that control the volatilization of an organic compound are solubility, molecular mass and structure, vapor pressure, Henry's law constant diffusivity, and the nature of the interface across which the organic must pass.

Volatilization rates from wastewater vary over a large range. Some chemicals volatilize from well-mixed water surfaces in a matter of hours, while others may remain in the water almost indefinitely unless they degrade or are removed by a different transfer mechanism.

Volatilization is affected by physical influences on the water or vapor interface—such as internally applied mixing energy—and air movement over the liquid or vapor interface. The latter can add mixing energy to the fluid and remove evaporated organic molecules from the liquid surface. Chemical properties of the aqueous phase, including the presence of modifying materials such as adsorbents, *colloids* (fine, suspended particles), organic films, *electrolytes* (conducting compounds), and *emulsions* (undissolved liquids), can influence volatilization. Volatilization tends to be temperature insensitive, as the principal effect of temperature is on vapor pressure.

OCTANOL–WATER PARTITION COEFFICIENT. The octanol–water partition coefficient (K_{ow}) is the ratio of the concentration of a chemical in the *n*-octanol phase to its concentration in the aqueous phase in an *n*-octanol or water mixture at room temperature. K_{ow} is measured with dilute solute concentrations (< 0.01 M) and varies little with solute concentration or with temperature. The octanol–water partition coefficient is measured by adding the test organic chemical to an *n*-octanol and water mixture, shaking the mixture for 15 minutes to 1 hour until equilibrium is reached, and centrifuging the mixture to separate the two phases. The chemical concentration in each phase is determined using an appropriate analytical technique.

The octanol–water partition coefficient is an important parameter in studies of the fate of organic chemicals in environmental systems. Measured values of K_{ow} span a large range, from 10^{-3} to 10^7. K_{ow} is inversely related to water solubility. Chemicals with low values of K_{ow} are *hydrophilic* (easily hydrated), have high water solubilities, and tend not to adsorb on organic matter. Chemicals with high values of K_{ow} tend to be hydrophobic. In wastewater treatment, organic chemicals with high K_{ow} are more likely to be removed by adsorbing onto other organic solids.

ADSORPTION COEFFICIENT. The adsorption coefficient (K_{oc}) is an equilibrium measure of the partitioning of a chemical between the solid and solution phases of either wastewater, a water-saturated or unsaturated soil, or of runoff water and sediment. K_{oc} is the ratio of the amount of chemical adsorbed per unit weight of organic carbon in the solid phase to the concentration of the chemical in the solution phase when the phases are at equilibrium. In soil-aqueous or sediment-aqueous systems, K_{oc} has been found to be a chemical specific adsorption parameter largely independent of soil or sediment properties, thus is determined by physical and chemical properties of the organic chemical.

The degree to which a chemical is adsorbed influences its partitioning to solids, its possible removal by solid or liquid separation processes (sedimentation, filtration, or centrifugation), and its removal or transformation by volatilization, photolysis, hydrolysis, and biodegradation. Values of the adsorption coefficient vary depending on the characteristics of the solid phase, particularly the organic content and the particle size distribution. Other factors influencing the value of K_{oc} are the pH, ionic strength, concentration of dissolved organic carbon in the water, and the cation exchange capacity of the soils. In addition, volatile chemicals may be lost because of volatilization, biodegradation, or adsorption onto the walls of the container.

ACID–BASE REACTIONS. Acid–base interactions between the chemical and the aqueous or solid phases in a reactor can exert a significant influence on partitioning. Moreover, an ionized organic acid or base can react differently than neutral molecules through its adsorption solubility and toxicity. Neutral species tend to be adsorbed to a greater degree than do ionized species. The extent to which chemicals are ionized in wastewater is a function of the solution pH and the acid dissociation constant.

RATE OF HYDROLYSIS. *Hydrolysis* is a chemical reaction in which water is added to organic molecules, cleaving a bond between carbon and another element in the original molecule. This forms a new bond between the carbon atom and the oxygen from water, with the element originally bonded to the carbon being displaced by hydroxide (OH^-). Hydrolysis is one of the most significant reactions in wastewater treatment and is an important initial step in biological degradation of organic materials present in nonsoluble form, such as the waste sludges treated in anaerobic digesters. Hydrolyic reactions act on many types of organic compounds. The rate of hydrolysis increases with temperature and is a function of the dissolved organic content, pH, and ionic strength of the medium.

When a compound undergoes hydrolysis, a *nucleophile* (electron donor such as water or a hydroxide ion), attacks an *electrophile* (electron receiver such as carbon or phosphorus) and displaces a group such as chloride or phenoxide. This reaction can be unimolecular or bimolecular. If it is a unimolecular process, it is characterized by a rate independent of the concentration and nature of the nucleophile. The bimolecular process depends on the concentration and identity of the nucleophile.

The hydrolysis rate depends partly on the products of the reaction and temperature. The decrease in concentration of a specific organic as a function of time is monitored by an appropriate method. For example, extraction and chromatographic analysis, ultraviolet light absorbance at a frequency characteristic of the organic, or determination of the concentration of the displaced element released.

ELIMINATION. *Elimination* is a chemical reaction in which two groups, such as hydrogen and chlorine (a halogen), are lost from adjacent carbon atoms so that a double bond is formed. For example, halogenated *alkanes* (carbon chains linked by single bonds) in an aqueous medium can undergo elimination reactions that produce *alkenes* (carbon chains with one double bond). In di-haloelimination, two halogens are removed from adjacent carbons. Elimination reactions can be competitive with hydrolysis for some types of organic compounds.

PHOTOLYSIS. *Photolysis* is a chemical transformation that results from an organic compound absorbing light energy. A compound can undergo photochemical reactions either directly, when the organic compound itself absorbs solar radiation, or through sensitized photolysis, where energy is transferred from another species in the solution. Photochemical reactions can result in many reactions in organic molecules such as fragmentation to *free radicals* (unpaired molecules) or neutral molecules, rearrangement and reactions, and electron-transfer reactions.

The rates at which photolysis reactions occur for a specific chemical in wastewater depend on the amount of incident solar radiation, the transparency of the aqueous medium, and the properties of the chemical. These reactions are only significant in uncovered tanks, ponds, and lagoons exposed to sunlight. However, the solar energy incident on the surface of an aqueous medium is not uniformly transmitted, and the chemical characteristics of the water influence the degree of penetration. Also, suspended sediments and surfactants can influence the rate and products of photochemical reactions. Thus, it is not currently feasible to predict the nature, rate or extent of photochemical transformations that occur during wastewater treatment.

BIOLOGICAL TRANSFORMATION. *Biological transformation* of organic compounds by microbial cells depends on many factors relating to the chemical structure and properties of the organic compound, its availability to organisms, the mass and type of the microbial population, and environmental factors. The species of organisms present in biological reactors dictate the types of reactions that can be mediated and the potential of the treatment system to transform specific organic compounds. The concentration of organisms and certain environmental conditions, such as temperature, pH, nutrients, and the availability of electron acceptors or donors, are important in controlling the effectiveness and efficiency of biodegradation. Factors that affect the availability of the organic compound to the organism, such as sorption and volatilization, also influence this process.

Biodegradation. *Biodegradation* refers to any structural change in a compound brought about by a biologically mediated process. Biodegradation requires:

- The proper organismal population with appropriate enzymes that can attack the compound;
- Proper environmental conditions for growth of the organisms and the functioning of its enzyme;
- That the compound and its bonds requiring cleavage be accessible to the organism and enzymes; and
- Enzyme induction if the required enzymes are not *constitutive* (immediately inducible).

Mineralization is the biodegradation of an organic compound to inorganic products. Under anaerobic conditions, the products of mineralization include carbon dioxide, water, nitrate, sulfate, chloride, and other elements or oxidized compounds, depending on the structure of the organic. Under anaerobic conditions, the products are methane, carbon dioxide, water, ammonia, hydrogen sulfide, chloride, or other elements or reduced compounds released from the organic substrate.

For any given organismal population and environment, organic compounds can be classified into several categories based on their receptiveness to biodegradation. Readily degradable compounds include those that are quickly degradable by constitutive enzymes—such as some sugars—and amino and fatty acids. For example, substrates of the central metabolic pathways of organisms and compounds that can enter intermediate stages of these pathways are readily degradable. Some anthropogenic compounds (such as chlorinated pesticides), naturally occurring humic substances, and lignin (a complex, woody plant fiber) degrade slowly or not at all. The classification of the degradability of a compound may depend on the types of organisms available and on the operation of the treatment system.

Aerobic and Anaerobic Processes. Biodegradation typically occurs through oxidative, reductive, or hydrolytic reactions. In oxidative or reductive reactions, electrons are transferred from electron donor compounds (which are oxidized) to electron acceptor compounds (which are reduced). Highly reduced organic compounds often are used as microbial substrates and oxidized; a highly oxidized compound has few remaining electrons available for transfer.

Biological processes are classified according to the final disposition of hydrogen (electrons) removed in oxidative reactions. The terminal electron acceptor strongly influences the types of biodegradative reactions that can or will occur. In aerobic processes, such as those that occur in activated sludge or aerated lagoon treatment, molecular oxygen is the terminal electron acceptor and readily degradable organic compounds are metabolized by bacterial cells to carbon dioxide and water. An example of a *catabolic* (energy producing) component of a biological reaction is the oxidation of glucose:

$$C_6H_{12}O_6 + 6O_2 \rightarrow 6CO_2 + 6H_2O \tag{8.3}$$

In anaerobic environments, such as in anaerobic digesters, organic compounds can be transformed by numerous complex reactions. Numerous fermentation reactions, in which no external electron acceptors are required, can occur in digesters. Fermentations result in the formation of many simple organic compounds (including organic acids and alcohols), most frequently from carbohydrates but also from amino acids and organic acids. Glucose transformation to ethanol and carbon dioxide is an example of a fermentation reaction:

$$C_6H_{12}O_6 \rightarrow 2(CH_3CH_2OH) + 2CO_2 \tag{8.4}$$

Propionate oxidation to acetate is an example of an anaerobic oxidation in which electrons are transferred to molecular hydrogen gas (H_2):

$$CH_3CH_2COO^- + 2H_2O \rightarrow CH_3COO^- + CO_2 + 3H_2 \tag{8.5}$$

A final reaction in anaerobic digesters is the reduction of carbon dioxide to methane:

$$CO_2 + 4H_2 \rightarrow CH_4 + 2H_2O \tag{8.6}$$

Degradation reactions and products of the major organic compounds found in wastewater are illustrated in Figure 8.4. Some pathways and reactions are similar in aerobic and anaerobic processes and involve similar transformations of organics. For example, hydrolysis of *colloidal* (suspended) organic material and other reactions by which polymers (complex molecules) or small molecules are prepared for entry to central biodegradative pathways proceed in the same manner in either the absence or presence of oxygen. Cen-

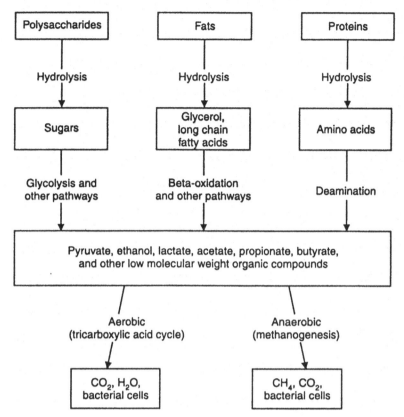

Figure 8.4 Degradation of major organic compounds found in wastewater.

tral catabolic pathways, including glycolysis and part of the tricarboxylic acid cycle, are used by both aerobic and anaerobic organisms.

Transformation of Xenobiotic Organics. Xenobiotic organic compounds contain chemical structures or substituents different from naturally occurring compounds. Xenobiotics are often recalcitrant (stubborn), remaining for an extended time in the environment either because they are intrinsically non-biodegradable or because conditions for biodegradation are not achieved. The chemical structure of the compounds, the type and position of substituents, the neighboring organismal population, and other environmental factors influence recalcitrance. Recalcitrant compounds may be nonbiodegradeable, slowly degraded, or selectively degraded, depending on microbiological and environmental conditions.

Bacteria attack xenobiotics through four major types of enzymatic reactions: oxidative, reductive, hydrolytic, and synthetic.

The chemical structure of a compound can influence biodegradability as a result of either specific physical properties that influence volatilization, sorption, and intrinsic biodegradability, or because of the presence of chemical substituents that are either inaccessible to or nonreactive with the available microbial enzymes.

Although there are no universally valid correlations between molecular structure and biodegradability of organic compounds, some typical relationships have been established. For example, chlorinated aliphatics are better degraded if the chlorine (⁻Cl) substituent is more than six carbon atoms from the terminal last carbon. Aromatic (ring) compounds are more easily degraded if they possess certain substituents (-OH, -COOH, -NH, or -OCH$_3$) than if they possess others (-F, -Cl, -NO$_2$, -CF$_3$, or -SO$_3$H). Furthermore, mono- and bicyclic aromatic hydrocarbons are more likely to be degraded than are polycyclic aromatic hydrocarbons; and alkanes (carbon chains with single bonds) having greater than 12 carbons are more readily degraded than those containing fewer carbon atoms.

Organisms are capable of degrading recalcitrant organic compounds to gain carbon or energy for growth. However, they sometimes degrade these compounds without deriving either carbon or energy through a process called *co-metabolism*. In co-metabolism, or secondary substrate use, an organism can transform an organic compound through one or a few steps while using another primary substrate to obtain carbon and energy. The significance of secondary substrate use is that, in conjunction with the growth of large quantities of bacteria on more readily usable substrates, appreciable biological degradation of recalcitrant organic molecules can be achieved.

Toxic organic compounds in wastewater treatment

Design of wastewater treatment systems is based primarily on the removal of readily degradable organic materials. The most significant removal mechanism for these compounds is biological transformation (biodegradation). For slowly biodegradable (recalcitrant) organic compounds, however, other removal mechanisms can compete with biodegradation and account for a significant fraction of the overall removal. While biological transformation may still predominate, the ultimate fate of organic compounds depends on their chemical characteristics and concentrations, as discussed previously. Organic compounds can attach to the solid particles in wastewater (*sorb*) and be removed by sedimentation.

Alternatively, volatile compounds can be released to the atmosphere through volatilization. Other fates of organic materials in wastewater include undergoing chemical reactions such as oxidation, reduction, or hydrolysis. Toxic organic compounds can be transformed to other more toxic species, which could inhibit other biochemical reactions. In other cases, toxic compounds have minimal effect on the operation of biological wastewater treatment systems or pass through the treatment plant unaltered.

PRIMARY TREATMENT. Primary treatment includes a process called primary sedimentation. The removal of toxic organic contaminants by this process depends on the solubility of the organic and its ability to partition onto solids that have settled from the wastewater. Because primary sedimentation takes place under conditions of low mixing energy, volatilization is limited. The removal of toxic contaminants in primary treatment can be a function of concentration of solids in the wastewater.

Volatile compounds have less tendency to sorb to solid particles than do semivolatile compounds; however, in laboratory studies of biological systems partitioning of organic solutes between the liquid phase and biological solids reached or approached equilibrium in fewer than 15 minutes (Weber *et al.*, 1987, and Blackburn, 1987). Toxic contaminants that bind to wastewater particles will then be associated with the primary sludge and can have an effect on the digestion process.

Modifications to primary treatment include the use of coagulant chemicals to improve particle removal and the use of filtration to remove particles remaining in the wastewater after sedimentation. The effectiveness of these processes for removal of toxic organics depends on the ability of the organics to react with the coagulant chemical or to associate with the wastewater particles.

SECONDARY TREATMENT. Analysis of the fate of recalcitrant organic chemicals in secondary treatment processes is complex as a result of many physical, chemical, and biological reactions occurring simultaneously. Overall removal is calculated as the sum of all competing removal mechanisms, including volatilization to the gas phase, biodegradation, and sorption onto solids.

In activated sludge reactors, toxic organics can react in the following ways. Volatile compounds can be transferred out of the liquid phase and into the gas phase, entering the atmosphere. The degree to which this transfer takes place depends on the Henry's law constant for the compound, the diffusivity of the compound, and the aeration system used. Less-volatile compounds can sorb to activated sludge flocs and be removed in the secondary clarifier. Biological metabolism and co-metabolism are removal mechanisms for degradable compounds.

Several mathematical models have been developed to describe the removal of trace organics in activated sludge systems, and to quantify the fractional removal by competing mechanisms (Baillod *et al.*, 1990). One such model is the General Fate Model (GFM) developed by Namkung and Rittmann (1986). The GFM considers the processes of biodegradation, volatilization, and sorption and has been used to predict successfully the overall volatile organic chemical removal rates from two wastewater treatment plants.

Measured, combined removal rates of 11 volatile organic chemicals were found to be 82 and 86% in the two plants and were closely matched by model simulations. The GFM predicted that sorption was an insignificant removal mechanism for all 11 compounds studied. According to model simulations, biodegradation accounted for removal of at least 95% of the influent concentrations of 4 of the 11 compounds. Volatilization removed from 19 to 83% of the influent aliphatics not degraded under aerobic conditions. One compound was not removed by either biodegradation or volatilization.

Weber *et al.* (1987) reported that effluent and offgas concentrations of poorly degradable organic compounds could be significantly reduced by the addition of 25 to 50 mg/L of powered, activated carbon.

R*EFERENCES*

Anthony, R.M., and Breinhurst, L.H. (1981) Determining maximum influent concentrations of priority pollutants for treatment plants. *J. Water Pollut. Control Fed.*, **53**, 1457.

Baillod, C., *et al.* (1990) Critical Evaluation of the State of WPCF Research Foundation Project 90-1, Alexandria, Va.

Barton, D.A. (1987) Intermedia Transport of Organic Compounds in Biological Wastewater Treatment Processes. *Environ. Prog.*, **6**, 4, 246.

Blackburn, J.W. (1987) Prediction of Organic Chemical Fates in Biological Treatment Systems. *Environ. Prog.*, **6**, 4, 217.

Namkung, E., and Rittmann, B.E. (1986) Estimating volatile organic compound emissions from publicly owned treatment works. *J. Water Pollut. Control Fed.*, **59**, 670.

Pellizari, E., and Little, L. (1980) Collection and Analysis of Purgeable Organics Emitted from Wastewater Treatment Plants. EPA-600/2-80-017, U.S. EPA, Washington, D.C.

Stumm, W., and Morgan, J.J. (1981) *Aquatic Chemistry.* John Wiley & Sons, New York, N.Y.

Weber, W.J., *et al.* (1987) Fate of Toxic Organic Compounds in Activated Sludge and Integrated PAC Systems. *Water Sci. Technol.*, **19**, 471.

S*UGGESTED READINGS*

American Public Health Association (1992) *Standard Methods for the Examination of Water and Wastewater.* 18th Ed., Washington, D.C.

Bauer, E.J., and McCarty, P.L. (1985) Utilization Rates of Trace Halogenated Organic Compounds in Acetate-Grown Biofilms. *Biotechnol. Bioeng.*, **27**, 1564.

Bell, J., *et al.* (1988) Investigation of Stripping of Volatile Organic Contaminants in Municipal Wastewater Treatment Systems. Ont. Ministry of the Environ., ISBN 0-7729-4720-1.

Bishop, D. (1982) The Role of Municipal Wastewater Treatment in Control of Toxics. Water Res. Div., Munic. Environ. Res. Lab., Cincinnati, Ohio.

Burns, G., and Roe, L. (1982) Fate of Priority Pollutants in Publicly Owned Treatment Works. Final Rep. 1 and 2, U.S. EPA, EPA-440/1-82-303, Washington, D.C.

Doull, J., *et al.* (1980) *Casarett and Doull's Toxicology, the Basic Science of Poisons.* Macmillan Publishing Co., New York, N.Y.

Gaudy, A.F., and Gaudy, E.T. (1980) *Microbiology for Environmental Scientists and Engineers.* McGraw-Hill, Inc., New York, N.Y.

Grady, C., (1985) Biodegradation: Its Measurement and Microbiological Basis. *Biotechnol. Bioeng.*, **27**, 660.

Hannah, S.A., *et al.* (1986) Comparative Removal of Toxic Pollutants by Six Wastewater Treatment Processes. *J. Water Pollut. Control Fed.*, **58**, 27.

Hsueh, K.P., and Wu, Y.C. (1989) Air Stripping of Volatile Organic Compounds Using Rotating Biological Contactors. *Proc. Int. Conf. Physicochem. Biol. Detoxification Haz. Wastes.* Technomic Publishing Co., Lancaster, Pa.

Johnson, L., and Young, J. (1983) Inhibition of Anaerobic Digestion by Organic Priority Pollutants. *J. Water Pollut. Control Fed.*, **55**, 1441.

Levine, A.D., *et al.* (1985) Characterization of the size distribution of contaminants in wastewater: treatment and reuse implications. *J. Water Pollut. Control Fed.*, **57**, 805.

McCarty, P., *et al.* (1981) Trace Organics in Groundwater. *Environ. Sci. Technol.*, **15**, 40.

Meyerhofer, J., *et al.* (1990) Control of Volatile Organic Compound Emissions During Preliminary and Primary Treatment. In *Report and Proceedings of the Air Toxic Emissions and POTWs Workshop.* Water Pollut. Control Fed./U.S. EPA, Washington, D.C.

Ng, A.S, and Stenstrom, M.K. (1987) Nitrification in Powdered-Activated Carbon-Activated Sludge Process. *J. Environ. Eng.*, **113**, 1285.

Owen, W.F., *et al.* (1979) Bioassay for Monitoring Biochemical Methane Potential and Anaerobic Toxicity. *Water Res.*, **13**, 484.

Parkin, G., and Speece, R. (1982) Modeling Toxicity in Methane Fermentation Systems. *J. Am. Soc. Chem. Eng.*, **108**, EE3, 515.

Parkin, G., *et al.* (1983) Response of Methane Fermentation Systems to Industrial Toxicants. *J. Water Pollut. Control Fed.*, **55**, 44.

Ribo, J., and Kaiser, K. (1987) Photobacterium Phosphoreum Toxicity Bioassay I. Test Procedures and Applications. In *Toxicity Assessment*, **2**, 305.

Rittmann, B.E. (1988) Potential for Treatment of Hazardous Organic Chemicals with Biological Processes. In *Biotreatment Systems.* D. Wise (Ed.), CRC Press, Boca Raton, Fla.

Russell, L., *et al.* (1983) Impact of Priority Pollutants on Publicly Owned Treatment Works Processes: A Literature Review. *Proc. Ind. Waste Conf., Purdue Univ.*, West Lafayette, Ind.

Stuckey, D., *et al.* (1980) Anaerobic Toxicity Evaluation by Batch and Semi-continuous Assays. *J. Water Pollut. Control Fed.*, **52**, 720.

Vasseur, P., *et al.* (1984) Luminescent Marine Bacteria in Ecotoxicity Screening Tests of Complex Effluents. In *Toxicity Screening Procedures Using Bacterial Systems*. D. Liu. and B. Dutka (Eds.), Marcel Dekker, Inc., New York, N.Y., 23.

Volskay, V., and Grady, C. (1989) Toxicity of Selected RCRA Compounds to Activated Sludge Microorganisms. *J. Water Pollut. Control Fed.*, **61**, 1850.

Walker, J.D. (1988) Effects of Chemicals on Microorganisms. *J. Water Pollut. Control Fed.*, **60**, 1106.

Walker, J.D. (1989) Effects of Chemicals on Microorganisms. *J. Water Pollut. Control Fed.*, **61**, 1077.

Water Pollution Control Federation (1986) *Removal of Hazardous Wastes in Wastewater Facilities—Halogenated Organics*. Manual of Practice No. FD-11, Alexandria, Va.

Water Pollution Control Federation (1980) *Wastewater Sampling for Process and Quality Control*. Manual of Practice No. OM-1, Washington, D.C.

Yang, J., and Speece, R. (1986) The Effect of Chloroform Toxicity on Methane Fermentation Systems. *Water Res.*, **20**, 1273.

Young, J., and Tabak, H. (1989) Screening Protocol for Assessing Toxicity of Organic Chemicals to Anaerobic Processes. *Am. Water Works Assoc./U.S. EPA Int. Symp. on Hazardous Waste Treat.*, Cincinnati, Ohio.

Young, L., and Rivera, M. (1985) Methanogenic Degradation of Four Phenolic Compounds. *Water Res.*, **19**, 1325.

Chapter 9
Bioaugmentation

Bioaugmentation is the process of adding commercial bacterial products to wastewater for enhancement of biological process efficiency.

PASTEURIZATION

Pasteurization is a term associated with the dairy industry. In the wine industry, heating fruit juices to 63°C (145°F) for 30 minutes can kill the yeast naturally associated with the fruits from which the juices were derived. Fermentation of pasteurized fruit juices, then, can be initiated by the addition of bottom sediments (that still contained living yeast cells) taken from good batches of wine.

STARTER "PURE" CULTURES

A pure culture contains a single species of organism isolated from a mixed population. Such pure cultures allow the development of starter cultures, which are used to replace the bottom sediments from fermented juices for starting fermentation of pasteurized fruit juices.

Microbial cultures now are used in the brewery, dairy, pharmaceutical, and many other industries. Knowledge of specific organisms when grown under well-defined conditions can provide useful results. For example, by using different starter cultures, the dairy industry can produce a variety of cheese products from the same initial milk product. The identity of the organisms in the starter culture determines the type of cheese produced.

WASTEWATER TREATMENT AS A BIOLOGICAL PROCESS

Along with other uses, microbial cultures can prove useful in wastewater treatment, though "well-defined conditions" rarely apply to wastewater treatment. Wastewater is a complex mixture of materials and requires the addition of a variety of bacterial cultures for treatment. Because the growth conditions of these cultures may vary with the conditions of the wastewater, the results of bioaugmentation are not as predictable.

BIOAUGMENTATION PRODUCTS FOR WASTEWATER TREATMENT

FORMS OF PRODUCTS. Dry powders and liquid suspensions are the most common forms of bioaugmentation products. Some products also are available in the form of moist pastes. Dry powder products are produced most frequently by air drying. *Lyophilization*, a process of freeze drying, also is used for producing dry powder products, though this process is not used frequently because of manufacturing costs.

Liquid suspensions are produced by growing bacteria in liquid media with the subsequent addition of stabilizing agents to preserve the bacterial viability in the liquid suspensions. Stabilization processes, including stabilizing agents, typically are patented or highly guarded as trade secrets.

The shelf life of any particular product should be considered carefully. Dry powders typically are more stable than liquid suspensions, but even these products have a shelf life that should not be exceeded. The shelf life of most liquid products typically ranges between 6 months and 2 years, whereas dry products normally are stable for several years.

Most bioaugmentation products should be stored between 4 and 32°C (40 and 90°F), with either temperature extreme being avoided if possible. Dry powder products absorb moisture readily and should be kept sealed. Moist pastes typically are stored under refrigeration, though bioaugmentation products—with the possible exception of lyophilized products—should be kept from freezing.

Although the majority of bioaugmentation products contain common soil bacteria similar to those in a garden, standard safe-handling procedures should be applied. The use of splash goggles, a face mask, and rubber gloves normally are sufficient for operator protection. However, the proper health-related information, which can be obtained from Material Safety Data Sheets for a specific product, should be consulted before using any product.

COMPOSITION OF PRODUCTS. The activity of a particular product depends on several factors, including the types, viability, and concentration of bacteria it contains. Furthermore, overall activity depends on the combined actions of all bacteria within the product.

There are two main factors that determine the activity of a particular species of bacterium: the genetic makeup of the bacterium and the environment in which it is located. Even if a living creature is genetically capable of a specific activity, the environment determines the extent to which the activity will occur.

The most effective bioaugmentation product is not simply one that contains the greatest concentration of bacteria, but one that contains bacteria having the greatest genetic diversity. However, because many bioaugmentation products are sold by weight or volume, a high concentration of genetically diverse bacteria also is desirable, from an economic standpoint. Typically, bioaugmentation products average from 0.1 to 1.0 billion viable bacteria per millilitre or per gram (per ounce).

Manufacturers of bioaugmentation products consider different species of bacteria (including soil bacteria) amongst their products, based on the bacteria they believe to be important in wastewater treatment. Two genera of bacteria commonly found in bioaugmentation products are *Pseudomonas* and *Bacillus*. *Pseudomonas* bacteria are included because of their genetic diversity and broad range of bacterial activities. Pseudomonads are useful in treating various types of industrial wastes.

Bacillus bacteria are capable of producing large quantities of exoenzymes that work outside of the bacterial cells to break down food particles otherwise too large to be taken up by the microscopic bacteria. The resulting smaller subunits of food particles become available to all bacteria in the immediate environment.

Because of the production of large amounts of exoenzymes by bacteria, bioaugmentation products may decrease sludge production (see Table 9.1). Supposedly, one pound of solids is produced for each pound of particulate biochemical oxygen demand (BOD) removed from wastewater, whereas one-half pound of solids is supposed to be produced per pound of soluble BOD removed. Therefore, in theory, sludge production in municipal wastewater

Table 9.1   Possible capabilities of bioaugmentation products.

Biodegradation of recalcitrant organic compounds

Control odor production in collection systems

Control grease accumulation in collection systems

Reduce grease accumulation on the surface of clarifiers

Reduce aeration requirements

Control filamentous bacterial growths

Control foam production in aeration tanks and digesters

Improve biochemical oxygen demand removal

Improve cold-weather performance

Resist shock loadings

Fast start-ups and recoveries from toxic shocks

Allow for attainment of nitrification

Decrease sludge production

Improve sludge dewatering

Improve solids settling in secondary clarifiers

Improve performance of aerobic digesters

Improve performance of anaerobic digesters

---

treatment plants can be decreased by approximately 50% if all particulate BOD is converted to soluble BOD.

USE OF PRODUCTS. Many manufacturers recommend that bioaugmentation products be used on a daily basis. Applications of 0.1 to 0.3 g/m^3 (1 to 3 lb/mil. gal) (or 1 to 3 mL/m^3 [1 to 3 gal/mil. gal]) of wastewater to be treated have been used at municipal plants. At industrial wastewater treatment plants, dosages are determined by organic and hydraulic loads because of the significant variations in the types and concentrations of organics. Dosages of one pound (or gallon) of product per 200 mg/L of BOD per million gallons of wastewater have been used in industrial applications.

Most manufacturers recommend start-up dosages significantly greater than daily maintenance dosages. Start-up dosages typically depend on specific applications and could involve product quantities ranging from 3 to 10 times more than daily maintenance dosages. Start-up periods also vary from several days to several weeks.

Liquid bioaugmentation products typically require no pretreatment before use, though dry products frequently are rehydrated. Though rehydration of dry products is not always necessary, it is recommended in applications involving short hydraulic detention times. Manufacturers' recommendations

---

should be followed for the rehydration process, which typically involves no more than soaking dry products in warm water for several hours. Once rehydrated, dry products are not stable. Thus, only the recommended daily dosage of the product should be rehydrated, and all of the rehydrated products should be used promptly.

The successful use of bioaugmentation products often involves adding them before arrival at the target site to make them active and acclimated to their wastewater environment before the intended application point. Thus, addition of a product to a primary clarifier influent (or to the collection system, if a wastewater treatment plant has no primaries) should produce better results than if the product were added directly to the secondary system. Also, adding a product to a collection system upstream of a grease-laden wet well should provide better grease removal than if the product were added directly to the wet well.

Bioaugmentation products typically are applied in the form of slug doses. However, continuous product addition using a chemical feed pump is recommended for collection system applications, most activated sludge processes involving a hydraulic overload, and most fixed-film processes, which include the use of trickling filters and rotating biological contactors. Methods and sites of bioaugmentation product addition to wastewater often determine the degree of product effectiveness.

SUPPLIERS OF PRODUCTS. Table 9.2 lists several questions that can be used as a checklist when contacting bioaugmentation manufacturers or distributors. A list such as this may prove helpful because unlike chemical products, which frequently provide fast reactions, biological products require more time to produce results. Thus, depending on the conditions in a particular wastewater treatment plant, a time period of 3 to 6 weeks or more may be needed to obtain desired results. During this time period, significant costs could accumulate while the success of the treatment would remain unknown.

COST-EFFECTIVENESS OF PRODUCTS. Use of bioaugmentation products can be an inexpensive means of enhancing biological treatment efficiency. On a short-term basis, operation and maintenance costs associated with their use often proves to be less expensive than capital costs. Nonetheless, meaningful cost-effective analyses should include cost comparisons of all available alternatives and long-term payments of capital improvements.

The long-term cost-effectiveness of the use of bioaugmentation products also should be investigated. Net savings may be attained through lowered electrical energy costs resulting from decreases in both biomass aeration and sludge processing.

Table 9.2   Questions related to bioaugmentation products.

Are safety, health, and handling instructions clearly defined?

What are the types and quantities of bacteria in the product?

What can this mixed population of bacteria accomplish under various environmental conditions?

Are assay procedures available for independent verification of types, quantities, and activities of bacteria within the product?

Are treatment proposals tailored to each particular plant or are instructions simply printed on the product label?

Are system analyses available on request before product usage and, if so, what are the costs of such analyses?

What is the time period required for attainment of results?

Can a monitoring program be established to determine product effectiveness even before attaining desired results?

What are the daily treatment costs, rather than product costs, on a per-pound or gallon basis?

If the product can produce specific results, under what conditions would use of the product no longer prove cost-effective?

What reasonable guarantees does the manufacturer offer regarding product effectiveness?

What is the quality and quantity of service available in support of product usage?

*W*ASTEWATER APPLICATIONS OF BIOAUGMENTATION PRODUCTS

Even if they fail to produce desired results, bioaugmentation products likely would do little harm to most wastewater treatment systems. The one exception is with anaerobic digestion.

Anaerobic digestion is composed of two distinctly different bacterial processes, commonly referred to as first stage and second stage. First-stage processes involve the breakdown of solids and the production of volatile acids. Second-stage processes involve the consumption of volatile acids by methanogenic bacteria coupled with the production of methane and carbon dioxide.

Use of a bioaugmentation product to increase the rate of the first-stage reaction of anaerobic digestion would yield an increase in the production of volatile acid. If the bioaugmentation product does not contain methanogenic bacteria, then this increased production could result in more volatile acids than can be consumed in the second-stage reaction; thus, care must be taken when using bioaugmentation products in anaerobic digesters.

ENVIRONMENTAL FACTORS AND BIOAUGMENTATION EFFICIENCY

As stated previously, genetics and environment determine bacterial activity. In nitrification, for example, two genera of bacteria genetically capable of converting ammonium (NH_4^+) to nitrite (NO_2) and nitrite to nitrate (NO_3) are *Nitrosomonas* and *Nitrobacter*, respectively. However, the mere presence of these two bacteria does not guarantee that nitrification will occur. Rather, the process depends on other conditions in a wastewater treatment plant.

Environmental conditions capable of affecting the nitrification process are presented in Table 9.3. A bioaugmentation product composed solely of *Nitrosomonas* and *Nitrobacter* can compensate to varying degrees for only the final three environmental conditions listed. No concentration of nitrifying bacteria will compensate for insufficient dissolved oxygen or alkalinity, and no nitrification could occur if ammonium is not present.

Table 9.3   Environmental conditions capable of affecting the nitrification process.

Ammonium (substrate) concentrations

Dissolved oxygen concentrations

Bicarbonate alkalinity concentrations

Soluble carbonaceous BOD concentrations

Nitrifying bacteria concentrations

Hydraulic detention times

Wastewater temperatures

There are two main considerations regarding the influence of environmental factors on bioaugmentation efficiency. First, bacteria supply the genetic potential for particular activities, and the environment will determine whether these activities can occur. Second, bacteria from bioaugmentation products can supply the potential for activities that can be completely absent from or occurring at low levels in wastewater treatment plants. However, their activities are still directly under environmental influence. Many bioaugmentation failures stem from the fact that environmental conditions are not sufficiently considered before initiating a bioaugmentation treatment program. Relevant environmental conditions to consider in the plant may include such factors as hydraulic detention time, mixed liquor suspended solids concentration, dissolved oxygen concentration, and wastewater temperature.

NATURAL SEEDING WITH INFLUENT WASTEWATER

Unlike municipal plants, industrial wastewater treatment plants are not preceded by extensive collection systems, thus natural seeding (addition of bacterial population by domestic wastewater) is minimal. Although natural seeding is more prevalent in municipal wastewater treatment plants, fecal coliforms are the predominant bacteria in raw municipal wastewater. Because of their low genetic diversity, these bacteria contribute little to effective wastewater treatment.

Those bacteria in raw influent that do contribute significantly to wastewater treatment are *saprophytic* (feed on nonliving organic material) soil bacteria. *Pseudomonas* and *Bacillus*, both being saprophytic soil bacteria, are the major bacterial genera found in many bioaugmentation products. Essentially all genera of bacteria included in bioaugmentation products, with the exception of nitrifiers, can be classified as saprophytic.

A prevalent misconception about bioaugmentation involves the intended purpose of the products themselves. The purpose of these products is not to establish a dominant biomass that displaces the biomass established through natural seeding. Rather, they are intended to augment the natural biomass, thus the term "bio"-augmentation.

NATURAL SELECTION AND ACCLIMATION

Given sufficient time, the bacteria most capable of biodegrading specific waste products are selected and establish a biomass in wastewater treatment plants. Bacteria that establish a biomass either already have the genetic potential to biodegrade specific organic compounds or they acquire the potential through changes in their normal genetic composition (by *mutation*).

Many bioaugmentation products contain soil bacteria that have been mutated. Mutation should not be confused with acclimation, which is simply a process whereby bacteria physiologically adjust to their environment. However, acclimation cannot occur if bacteria do not already have or subsequently acquire the genetic potential to adjust to environmental conditions.

Under natural conditions, mutations occur at low levels. They can be accelerated through the application of specific techniques, resulting in

- No apparent changed in microbial activities;
- Elimination of a specific activity;
- Reduction of a specific activity;
- Enhancement of a specific activity; or
- Generation of a totally new activity (rarely).

The enhancement of a specific activity is the main thing manufacturers of bioaugmentation products look for when isolating mutated bacteria. Bacteria that are mutated typically are prescreened for a low-level ability to biodegrade a specific compound, or group of compounds. The mutated bacteria are then further screened to determine if any have enhanced abilities to biodegrade the compound or group of compounds being investigated. If such a mutated bacterium is found, it is isolated and included in a specific bioaugmentation product.

Another benefit provided by mutating bacteria is that specific activities sometimes can be released from environmental control. In other words, a specific bacterial activity might be able to occur independently of environmental conditions, thus more constantly. Isolating specific bacterial activities can contribute to effective treatment of some industrial wastes. Although applicable to municipal wastewater treatment, bioaugmentation products comprised of mutated bacteria typically are used for the treatment of industrial wastes.

The process of bacterial mutation differs from creating genetically engineered bacteria. A genetically engineered bacterium typically contains a new genetic potential acquired from foreign sources. None of the currently available bioaugmentation products contain bacteria created from genetic engineering. It is illegal to release genetically engineered bacteria to the environment without prior approval from the U.S. Environmental Protection Agency.

One of the premises of bioaugmentation is that natural selection, or "survival of the fittest," does not always result in optimum wastewater treatment. A bacterium capable of maximum use of a specific organic compound is not, depending on environmental conditions, necessarily the "most fit" bacterium or the one most capable of reproducing. Another bacterium capable of using the same organic compound, but at one-tenth of the maximum rate, could become the dominant component of a biomass if environmental conditions provide a reproductive advantage for even a few minutes. Domination of the biomass by such a less- capable bacterium would result in a wastewater treatment plant having greatly reduced organic loading capabilities.

*O*PERATIONAL CONTROL AND BIOAUGMENTATION

Questions frequently arise as to whether results attained from bioaugmentation programs result from the products themselves or from more attentive operational control. Results obtained from the use of bioaugmentation products depend on both product effectiveness and effective operational control.

Effective operational control can be measured in terms of an effective biomass. When such a biomass is lacking, so is effective operational control, regardless of the capabilities of an operator. The effectiveness of any particular bioaugmentation product should be judged by the relative degree that the product allows an operator to control the biomass, which subsequently results in more effective wastewater treatment.

TERMINATION OF BIOAUGMENTATION AND SUBSEQUENT TREATMENT RESULTS

To maintain favorable results, manufacturers recommend continued use of bioaugmentation products, based on the process of natural selection, which may operate against a beneficial bacterium whether it is placed in a wastewater treatment plant through natural seeding or through bioaugmentation. Such a bacterium can eventually be lost if it is not fit enough to compete and reproduce in the biomass environment.

For this reason, manufacturers recommend significantly greater start-up dosages of bioaugmentation products to first establish the bioaugmentation bacteria as integral components of a natural biomass. The theory behind the need for continued daily supplements is that those beneficial bacteria lost through natural selection must be replaced to maintain their populations at optimum levels.

Once a biomass has been established in a wastewater treatment plant, at least one full mean cell residence time theoretically is required before any subsequent changeover in biomass composition. Depending on conditions existing in a wastewater treatment plant, termination of a bioaugmentation program typically requires one-half to two mean cell residence times for a complete loss of activities that might have resulted from the program. The termination of bioaugmentation programs have not been found to result in lower treatment efficiencies than were being attained before initiation of the programs.

Index

A

Activated sludge, 5, 129
Adsorption coefficient, 161
Amino acids, 27
Anaerobic bacteria, 93, 94
Anaerobic digestion, 7, 95, 96
  mesophilic, 99
  psychrophilic, 99
  thermophilic, 100
  toxic substances, 101
    cyanide, 105
    dissolved ammonia gas, 103
    heavy metal ions, 101
    sulfide inhibition, 102
    un-ionized volatile acids, 104
Anaerobiosis and methanogenesis, 97
  dissolved oxygen, 98
  nutrients, 97
  pH, 100
  temperature
    mesophilic anaerobic
      digestion, 99
    psychrophilic anaerobic
      digestion, 99
    thermophilic anaerobic
      digestion, 100
  volatile acids, 101
  volatile solids, 101
Autotrophs and heterotrophs, 79
  carbon cycle, 80
  corrosion, 82
  interactions, 79
  nitrogen fixation, 80
  nitrogen cycle, 80
  sulfur cycle, 80

Autotrophs, wastewater treatment, 76

B

Bioaugmentation, 171
  efficiency, 177
  products, 172-175, 177, 179
    applications, 176
    cost, 175
    wastewater applications, 176
  natural seeding, 178
  natural selection/acclimation, 178
  operational control, 179
  pasteurization, 171
  starter cultures, 171

C

Carbon cycle, 80, 81
Carbon removal, 58
Chromatography, 155-157

E

Enzymes, 11, 13
  cellular location, 12
  characteristics, 11
  environmental change, 13

---

solubility, 158
vapor pressure, 160
volatility, 160
classification, 142
molecular size, 148
quantification, 149-158
biochemical oxygen demand, 150
chemical oxygen demand, 153
chromatography, 155
detection devices, 156
gas chromatography, 155, 156
liquid chromatography, 157
mass spectrometry, 156
theoretical oxygen demand, 153
thin-layer plate chromatography, 158
total organic carbon, 153
total organic halogen, 154
ultraviolet absorption, 154
volatile solids, 154
toxic organic chemicals, 141
aliphatic alcohols, 146
aliphatic hydrocarbons, 145
aromatic hydrocarbons, 146
halogenated aliphatic hydrocarbons, 146
halogenated aromatic hydrocarbons, 146
toxicity, 166, 167
wastewater treatment, 146
Oxidation of sulfur and iron compounds, 79

P

Pasteurization, 171
Phosphorus removal, 59
Photosynthetic organisms, 84
aerobic algae ponds, 85
algae, 84
anaerobic ponds, 85
cyanobacteria, 84
facultative ponds, 85
green bacteria, 84

performance, 85
purple bacteria, 84
Phototrophic organisms, 83
Purines, 27
Pyrimidines, 27

S

Secondary treatment, 3
activated sludge, 5
oxidation ponds, 6
trickling filtration, 3
Sludges, 7
aerobic digestion, 7
anaerobic digestion, 7
bulking, 41, 73
foaming, 43, 73
Starter "pure" cultures, 171
Sulfur cycle, 80, 82

T

Toxic organic compounds, 145-146, 166
primary treatment, 167
secondary treatment, 167
Trace element removal, 61

V

Vitamins, 27
Volatile acids, 89
carbohydrates, 92
lipids, 90
polysaccharides, 92
proteins, 92